THE STEVIA DECEPTION

The Hidden Dangers of Low-Calorie Sweeteners

Dr. Bruce Fife

Piccadilly Books, Ltd.
Colorado Springs, CO

Every effort has been made to ensure that the information contained in this book is complete and accurate. However, neither the publisher nor the author is engaged in rendering professional advice or services to the individual reader. The ideas and suggestions contained in this book are not intended as a substitute for consulting with your physician. Neither the publisher nor the author are responsible for your specific health or allergy needs.

Piccadilly Books, Ltd.
P.O. Box 25203
Colorado Springs, CO 80936, USA
info@piccadillybooks.com
www.piccadillybooks.com

Library of Congress Cataloging-in-Publication Data

Names: Fife, Bruce, 1952- author.
Title: The stevia deception : the hidden dangers of low-calorie sweeteners / by Dr. Bruce Fife.
Description: Colorado Springs, CO : Piccadilly Books, Ltd., [2017] | Includes bibliographical references and index.
Identifiers: LCCN 2016037951 | ISBN 9781936709113 (pbk.)
Subjects: LCSH: Sugars--Health aspects. | Sweeteners--Health aspects. | Sugar-free diet. | Stevia rebaudiana.
Classification: LCC QP702.S85 F54 2017 | DDC 612/.01578--dc23 LC record available at https://lccn.loc.gov/2016037951

Printed in the USA

Contents

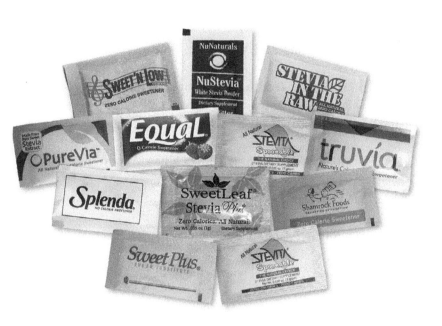

The Bittersweet Truth About Stevia

IF I HAD ONLY KNOWN

"I am experiencing a quite frightening health situation and I can't help but believe that stevia has something to do with it," says Tammy, a 42-year-old homemaker. "I quit using stevia about two weeks ago, after using it for four years, and now I'm experiencing dizzy spells and loss of balance after eating anything with even the smallest amount of sugar in it. I can eat fruit, but not sugar. Needless to say, I'm avoiding all sugar."

Tammy's doctor didn't know what was wrong and suggested she might have a blood sugar problem, which was surprising to her since she didn't have any blood sugar issues before she went off sugar and started using stevia. Was it possible that long-term use of stevia made her glucose-intolerant?

"I first started using [stevia] before a pregnancy," Tammy says. "I wanted to use a sweetener that would be safe for the baby. I ended up having a miscarriage." Interestingly, spontaneous abortion is one of the documented, but little known, side effects associated with stevia. On some packages of stevia you will even find a warning statement that reads: "Caution: *not recommended for pregnant women*, children or those who have low blood pressure. Keep out of reach of children" (italics added). It may also warn: "Do not exceed five packets per day." If stevia is made from a harmless plant, why the warnings?

"I continued to use stevia," says Tammy, "because I thought it was healthier than all alternatives. I could find *nothing* saying there were any dangers in using it, except for those who are on blood pressure medications, because it can cause blood pressure to decrease.

"I had stevia about three to four times a day, and worked up to a dropperful each time. Very shortly after starting to use stevia, I'd say about two weeks, I started to develop horrific gas and constipation, and had a very difficult time losing weight after losing the baby. At the time, I attributed these symptoms to pregnancy and post-pregnancy. I've been to countless specialists and had numerous unpleasant gastro tests performed, all with normal results. I never dreamed stevia was the culprit. After all, it's natural and supposedly has *no* side effects, right?

"Along with the gastro issues I was constantly exhausted and had typical hypothyroid symptoms. I started seeing my endocrinologist, and he prescribed Armour Thyroid."

Tammy began an exercise program to help her lose the extra weight she had gained during pregnancy. She took brisk walks daily and joined a Pilates class that used light hand weights. It wasn't easy, but in time she dropped down to 147 pounds (66.6 kg). "After every workout I felt so drained that I either wanted to throw up or pass out. A few times I felt myself about to faint, but was able to stop it. So, on the advice of my instructor, I started adding more protein to my diet. This didn't really help. The only thing that made me have strong workouts was eating a sizable lunch of rice and veggies, but the amount of food that made me have a good workout (without leaving me feeling like I could pass out), made me start gaining weight. So I cut back on the food and resumed weak workouts. I began to make a point of noting what I ate and drank when I felt the worst. At first, I thought it was coffee. Then I thought it was tea. Then I realized I felt worst after a workout where I had stevia in my drink with my meal before working out."

Realizing that stevia might be what was causing her problem, she stopped using it and replaced it with a little sugar. But that brought on new symptoms. "I have now been off of stevia for about two weeks," she says, "and now I am experiencing scary symptoms. I am having dizzy spells. They come after I've had food, drink, or gum that contains *any* amount of sugar."

Despite the apparent discovery of glucose intolerance, her overall health improved. "Since I stopped using stevia my digestive system has vastly improved to what I can happily call normal. A few other aggravating issues are gone now too, like sharp pains in my breasts and strong body odor from my right armpit only (that would reappear shortly after taking a shower, and was completely resistant to deodorants and antiperspirants). I feel great if I just stay away from sugar. Pasta doesn't bother me and neither does rice or carby stuff. I can have fruit and even wine with no dizziness. I've started using Splenda out of desperation . . . I am not willing to restart using stevia. The side effects I would care not to live with again."

Most people are totally unaware of the possible side effects associated with stevia. We are so inundated with the propaganda that stevia is "natural," "an herb," "harmless," even "healthy," that we are brainwashed into believing it. If anyone says otherwise, our first reaction is disbelief and maybe even anger or indignation. If Tammy had only known of the dangers of using stevia, she might not have lost her baby, or experienced four years of gastrointestinal discomfort, breast pain, fatigue, and weight problems, or developed blood sugar issues. The purpose of this book is to present to you the complete story of stevia, and provide you with the information you need to make an informed decision about whether or not stevia is right for you, so that someday you won't say, "If I had only known."

THE STEVIA MYTH

Refined sugar has long been linked to a multitude of health problems ranging from tooth decay and obesity to heart disease and diabetes. Most "natural" sweeteners such as honey, dehydrated sugarcane juice, maple syrup, and date sugar are perceived as slightly better because they are less processed and retain a tiny amount of vitamins and minerals. But even these so-called natural sweeteners deliver the same number of calories and can have the same effect on blood sugar and metabolism.

Low-calorie sweeteners have become popular as alternatives to sugar. At first, they seemed like a dream come true. They contain essentially no calories, so there is nothing to be converted into fat, and because they don't contain any actual sugar, they don't raise blood sugar levels—a major concern for diabetics. Low-calorie sweeteners give weight-conscious individuals, diabetics, and others the freedom to eat, without worry, the same sweet-tasting foods and beverages they have always enjoyed. But these low-calorie sweeteners are not without their own problems. They, too, can promote obesity and diabetes, in addition to numerous other health problems ranging from cancer to seizures, making them even worse than sugar.

For many years there just didn't seem to be a sweetener available that was completely harmless. Then it was discovered that the leaves of a little-known South American shrub called stevia (*Stevia rebaudiana Bertoni*) could be used as a low-calorie sweetener. Stevia has a sweet taste but contains no sugar and essentially no calories. It is appealing because it is derived from a plant, and, thus is perceived to be more natural and less harmful than artificial sweeteners concocted in a chemist's laboratory. Here is a natural sweetener that appears to have none of the detrimental effects associated with sugar or artificial sweeteners. It immediately became popular as a non-caloric sweetener among health-conscious individuals and others who want to reduce their sugar and carbohydrate intake. Currently, hundreds of products are sweetened using stevia extract.

Stevia comes from a small shrub native to Paraguay and Brazil, where it is known as the "sweet herb." Its leaves have a sweetness about 30 times greater than sugar. The Guarani Indians, who live in the region, have been using the herb for centuries as both a sweetener and a medicine. It is used to sweeten beverages, and reportedly to disinfect wounds and improve digestion. Its sweetening effect is well-known, its purported medicinal properties, however, have not yet been substantiated by medical research.

Ground or whole stevia leaf can be used to sweeten teas and strong beverages. Stevia used in leaf form is not practical for most other culinary uses because it tastes too much like an herb or a weed. A more useable form is stevia extract, which is a purified concentrate of steviol glycosides, the chemicals that give the plant its sweetness. Stevia extract is 150 to 200 times sweeter than sugar and does not have an herbal taste; other than the sweetness, it is essentially tasteless. The extract is available as a powder or liquid.

Because of its sweetness, only a small amount is needed to sweeten foods or drinks.

If you use too much, however, it produces a bitter, molasses-like aftertaste, and for this reason many people don't like it. Even a tiny amount can be too much; it takes practice to use just enough to slightly sweeten foods without getting the strong aftertaste.

Like so many other people, I was enamored by the idea of a harmless, herb-based sweetener. I used stevia, my family used it, and I recommended it to others. I wrote about it in some of my books as an alternative to sugar and other sweeteners, and even developed a number of recipes using it. But time and time again I observed undesirable side effects. At first I brushed them aside without a second thought; I wanted to believe that stevia was good for me, or at least better than sugar. I perceived it to be harmless; after all, it was derived from an herb, so how bad could it be, right? I wanted to believe this so badly I ignored evidence that suggested otherwise.

One of the undesirable effects I observed was addiction. Stevia appeared to be just as addictive as sugar. People who were addicted to sugar when they began using stevia simply switched their addiction from sugar to stevia. They would eat the same unhealthy foods they had before, but just sweetened them with stevia instead of sugar.

Another problem was that people using stevia had a very difficult time losing weight. Stevia seemed to block weight loss even when combined with very strict weight loss diets. It was more of an anti-diet product.

One of the most troubling effects that I encountered with stevia was that it would prevent nutritional ketosis—a goal of a therapeutic ketogenic diet. I have written a number of books on the therapeutic benefits of the ketogenic diet. The ketogenic diet has long been used in the treatment of epilepsy, and is currently being used in the treatment of a variety of neurological disorders ranging from Alzheimer's to autism.[1-3] It has become a popular treatment for obesity and has proven to be an effective means of losing excess body fat and reducing weight.[4] World-class and amateur athletes are now beginning to use the ketogenic diet to boost their performance in practice and competition.[5] I often recommend the diet for various reasons, and have observed that when people on the diet were using stevia it would prevent them from getting into ketosis. If they were already in ketosis before using stevia, it would kick them out in a snap. In short, stevia appeared to be anti-ketogenic—which was disturbing because a ketogenic diet can be very therapeutic for many health problems.

As I continued to use and recommend stevia, I kept seeing these disturbing side effects, to the point that I could no longer keep ignoring them. I had to find out what was going on. I decided to investigate stevia thoroughly and learn everything I could about it. I looked up and read every study I could find on stevia and low-calorie sweeteners. What I discovered shocked me! I learned why I was seeing these unpleasant side effects. I also discovered that the stevia we use as a sweetener and that is used in food manufacturing was more like an artificial sweetener than it was an herb. It all began to make sense.

I am reminded of another product that was once held in high esteem as a healthy natural product but later was found to cause serious health issues. For many years, we believed that soy was a health food. It does have many positive attributes: it is a good source of plant-based protein, has a number of beneficial vitamins and minerals, and contains phytoestrogens, which have been used to treat hormone problems in women. Those people and industries promoting and selling soy ignored the negative effects and loudly proclaimed the virtues of their products. Medical professionals and researchers convinced of soy's benefits wrote articles and published studies to further the myth. However, soy has many undesirable effects as well. For instance, the phytoestrogens that provide a pharmacological benefit can also stir up a great deal of harm. Anti-nutrients in soy block nutrient absorption. Soy goitrogens can cause hypothyroidism. The fat in soybeans is highly polyunsaturated and can promote many health problems. Indeed, there are entire books written on the dangers of consuming soy.[6]

Although artificial sweeteners have received their share of criticism, stevia is looked upon as the darling of the health food industry. It is promoted as an herbal sweetener that not only is harmless but also provides numerous health benefits. This wholesome image, however, conceals a dark side. I was fooled, and although I witnessed some problems with stevia I refused to believe there was anything fundamentally wrong with it—until, that is, I read the research and saw the evidence. Once I learned the truth, I stopped recommending stevia as a substitute for sugar, and decided to share this knowledge with others. That is how I came about writing this book.

We've been so misled about the true character of stevia that some people will find the information in this book difficult to accept. Please keep in mind, that I didn't make up the data in this book; I'm just reporting the facts gleaned from published medical and nutritional journals. Every statement I make that might be considered controversial is backed by references to these sources and by statements from experts. I'm giving you nothing but the facts, so that you can make an informed choice whether to use stevia or not.

IS IT AN HERB OR A DRUG?

The biggest misconception about the sweetener stevia is that it is a natural product—an herb or herbal extract—and therefore healthier than other low-calorie sweeteners. An advertisement promoting the use of Truvia-brand stevia sweetener as a better choice than Splenda sums up the general perception of stevia:

Is stevia sweetener an herb or a drug?

11

"Our sweetener is more than splendid, it's natural. Born from a leaf, not in a lab, our perfect sweetness comes from the leaves of the stevia plant. We just give them water. We give them sun. Next we steep them in a process like making tea. Ultimately these stevia leaves give back a recipe for sweetness that's both delicious and zero-calorie. How beyond splendid is that?" If you believe this advertisement, you, like most other people, have been deceived. The statements in this advertisement, while often repeated and widely believed, are pure fiction.

Marketers promote stevia as an "herbal" sweetener, implying that it is simply an herb and, therefore, safe and possibly even healthful. Dubious health benefits are often attributed to it to suggest its possible (but unproven) medicinal value. Stevia, as it is generally sold in stores, however, is no more natural than white sugar—or cocaine for that matter. The herb itself is used only to sweeten tea, it is never used as an all-purpose sweetener because of its strong alfalfa-like taste. Stevia extract is the product you purchase in the store for general use as a sweetener. It is a highly processed and refined chemical composed of pure steviol glycosides—the substances responsible for the sweet taste. The main steviol glycosides are stevioside and rebaudioside A. All of the stevia sweeteners you buy in the store contain purified steviol glycosides, primarily rebaudioside A, which has the greatest sweetening effect with the least bitter aftertaste. Rebaudioside A is also known as Reb A or rebiana.

When you extract plant chemicals, purify them, and consume them in concentrated doses, they can exert pronounced physiological effects. That is how we've gotten many of our drugs. Aspirin is an extract of willow bark; the heart drug digitalis is derived from the herb foxglove; cocaine comes from the leaf of the coca plant; and sugar is extracted from sugarcane and sugar beets. The stevia you purchase as a sweetener is no different. The extracted and refined rebaudioside A sold as "stevia" can no longer be considered an "herb" any more than the extract of beets (i.e., white sugar) can be considered a vegetable. To call stevia an herb or herbal sweetener is like calling sugar a vegetable or a vegetable sweetener.

Stevia extract (concentrated rebaudioside A) isn't a simple extract like fruit juice or tea, but a highly processed, refined, and purified crystalline powder, just like sugar or cocaine. Obviously, it cannot truthfully be considered an herb. It cannot be considered a food either, because it provides no nutritional value: it is not a carbohydrate, fat, or protein, and supplies no calories. It is not a nutritional supplement, as it provides no vitamins, minerals, or other essential nutrients required by the body.

So if stevia extract is none of these things, what exactly is it? When you get right down to it, it resembles a drug more than anything else. And as with other drugs, only a tiny amount, a few drops, are needed to cause pronounced alterations in normal physiology. The most obvious effect is its stimulation of sweet taste, with a sweetness intensity similar to other artificial sweeteners that possess drug-like qualities. Stevia alters normal homeostatic processes involved in appetite control, metabolism, and energy balance. Like many other drugs, it can become highly addictive.

STEVIOL GLYCOSIDES

Like all other plants, stevia leaf contains a mixture of vitamins, minerals, carotenoids, flavonoids, and other materials, some of which provide nutritional benefits. The stevia sweeteners you get at the store, however, have no nutritional value. All of these nutrients have been stripped away, leaving concentrated steviol glycosides.

Stevia leaf contains over 32 steviol glycosides in various concentrations.[7] The most predominate one is stevioside, constituting between 4 and 13 percent of the dry weight of the leaf. Rebaudioside A is the second most abundant glycoside, constituting 2 to 4 percent. Most of the rest make up very small or trace amounts. Each glycoside has a different sweetening potency.

Most stevia sweeteners are at least 95 percent, and as high as 98 percent, pure rebaudioside A; the remaining portion consists mostly of stevioside, with just trace amounts of the others. Some stevia sweeteners sold in health food stores are a little less refined, but generally contain at least 80 percent mixture of glycosides, predominantly stevioside.

Rebaudioside M has the greatest sweetening power, but is only found in trace amounts.

Steviol Glycosides		
Steviol Glycoside	**Sweetness Relative to Sucrose**	**Percent (dry wt) in Stevia Leaf**
Rebaudioside A	200	2-4
Rebaudioside B	150	trace
Rebaudioside C	30	1-2
Rebaudioside D	221	trace
Rebaudioside E	174	trace
Rebaudioside F	200	trace
Rebaudioside M	250	trace
Stevioside	210	4-13
Steviolbioside	90	trace
Rubusoside	114	trace
Dulcoside A	30	0.4-0.7

Sources:
Prakash, I, et al. Development of next generation stevia sweetener: Rebaudioside M. *Foods* 2014;3:162-175.
Kinghorn, AD and Soejarto, DD.1991. Stevioside. In O'Brien, NL. and Gelardi, RC. (Eds.). *Alternative Sweeteners*. New York; Basel; Hong Kong: Marcell Dekker Inc. p. 157-171.

HOW OUR BODIES PROCESS STEVIOL GLYCOSIDES

Steviol glycosides are actually a form of alcohol. Stevioside consists of a steviol molecule (a diterpenic carboxylic alcohol) attached to three glucose molecules. Rebaudioside A is almost identical to stevioside but contains four glucose molecules. The other steviol glycosides have similar structures.

Steviol glycosides are not well-digested by the body. The human digestive tract does not produce the enzymes necessary to break the chemical bonds; nor can digestive stomach acids break them. Bacteria in the small intestine also have no effect on these compounds. Consequently, steviol glycosides pass through the stomach and upper intestinal tract unchanged. This is why they provide no calories, no energy, and no nutritional benefit.

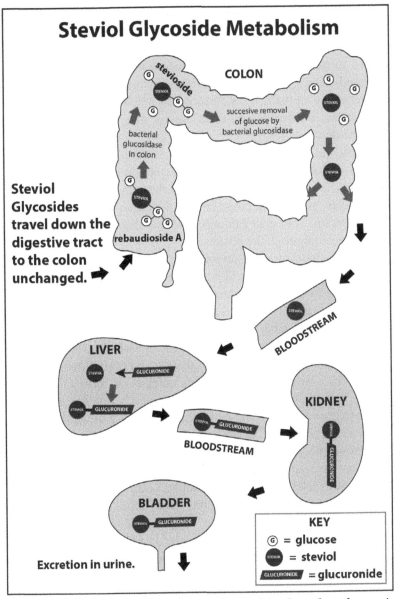

Steviol Glycoside Metabolism

COLON

stevioside

succesive removal of glucose by bacterial glucosidase

bacterial glucosidase in colon

Steviol Glycosides travel down the digestive tract to the colon unchanged.

rebaudioside A

BLOODSTREAM

LIVER

STEVIOL ← GLUCURONIDE

STEVIOL — GLUCURONIDE

BLOODSTREAM

STEVIOL GLUCURONIDE

KIDNEY

STEVIOL GLUCURONIDE

BLADDER

STEVIOL — GLUCURONIDE

Excretion in urine.

KEY
- Ⓖ = glucose
- STEVIOL = steviol
- GLUCURONIDE = glucuronide

Steviol glycosides enter the colon unchanged. In the colon glucose is removed from the steviol molecule. Steviol is then absorbed into the bloodstream and sent to the liver where it is combined with glucuronide and released back into the bloodstream. Steviol glucuronide is picked up by the kidneys and excreted in the urine.

When they reach the last segment of the intestinal tract— the colon—resident bacteria are capable of partially dismantling them by cleaving off the glucose molecules, reducing them down to steviol and glucose. As glucose molecules are removed, Rebaudioside A is transformed into stevioside and one free glucose molecule; eventually it is reduced to steviol and four glucose molecules. Stevioside and other steviol glycosides are similarly broken down.

Steviol is not broken down any further in the colon. The released glucose feeds the resident bacteria in the colon and is not absorbed into the bloodstream. With the removal of the glucose molecules, steviol can pass through the colon wall and enter the bloodstream. The small amounts of steviol that remain in the colon are excreted through the bowels.

In the bloodstream, steviol provides no useful purpose. It is a foreign body, a toxin, just like other forms of alcohol, and must be ushered to the liver and kidneys for rapid removal and excretion.

While stevioside and rebaudioside A appear to be nontoxic, steviol, on the other hand, has been shown to be toxic and mutagenic in lab animals.[8] At high doses steviol is toxic to pregnant hamsters and their fetuses when given on days six through ten of gestation.[9] University of Arizona toxicologist Dr. Ryan J. Huxtable says, "Steviol has the ability to penetrate cells, and has a number of potent effects on basic cell functions. The most important of these are inhibition of monosaccharide transport, and inhibition of oxidative phosphorylation, combined with disruption of a number of mitochondrial activities. In addition, steviol is bioactivated to a mutagenic metabolite, [steviol-16alpha,17-epoxide]."[10]

Steviol is transported to the liver to clear it out of the bloodstream. Here it is bound to glucuronide to form steviol glucuronide. The liver combines glucuronide to various toxins and drugs to reduce their toxicity and assist in their excretion from the body. The newly bonded steviol glucuronide is released into the blood. The presence of steviol glucuronide causes the blood vessels to dilate, which increases blood flow through the kidneys for rapid removal. In this process, urine flow is increased, which is the reason for the diuretic effect and the subsequent drop in blood pressure often reported in stevia studies. Steviol glucuronide is filtered out of the blood by the kidneys and exits in the urine.

LOW-CALORIE SWEETENERS

This book is primarily about stevia, but we can't discuss research on stevia without talking about the other sugar substitutes. For this reason, we need to define some terms that are used throughout this book and in the medical literature.

The term *low-calorie* sweetener refers to any sugar substitute that provides fewer calories than sugar. This would include artificial sweeteners, such as aspartame and sucralose; sugar alcohols, such as xylitol and erythritol; and plant-derived sweeteners, such as stevia and luo han guo.

Low-calorie sweeteners are classified into two categories. The first category includes sweeteners that provide no nutritional value or no calories. These are referred to as *nonnutritive, non-caloric*, or *zero-calorie* sweeteners. These sweeteners are also much sweeter than sugar, so they are also called *high-intensity* or *intense* sweeteners. Sweeteners in this category that are allowed in the United States by the Food and Drug Administration

High-Intensity Sweeteners	
Sweeteners	**Sweetness Relative to Sucrose**
Stevia leaf	30-40
Cyclamate	30-50
Stevia (mixed steviol glycosides)	150-200
Aspartame	200
Acesulfame K	200
Stevia (rebaudioside A)	200-300
Saccharin	300
Neohesperidin dihydrochalcone	340
Luo han guo (mogroside 5)	300-400
Sucralose	600
Thaumatin	2,000
Alitame	2,000
Neotame	8,000
Advantame	20,000
Sucrose has a comparative sweetness of 1	

(FDA) include saccharin, acesulfame K, aspartame, sucralose, neotame, advantame, stevia (steviol glycosides), and luo han guo. Other sweeteners in this category such as cyclamate, thaumatin, neohesperidin dihydrochalcone, and alitame, are authorized for use in some other countries. Most of these nonnutritive, high-intensity sweeteners are created in a laboratory, and are often referred to as *artificial* sweeteners. The American Diabetes Association prefers the term nonnutritive sweeteners, but you will see all of these terms used.

The second category of low-calorie sweetener is *nutritive* or *bulk* sweeteners. These are sweeteners that provide calories, but not as much as sugar. Sugar alcohols make up this category. Among the sugar alcohols, the FDA has approved the use of erythritol, hydrogenated starch hydrolysates (HSH), isomalt, lactitol, maltitol, mannitol, sorbitol, and xylitol. The European Commission has approved all of these as well as polyglycitol syrup.

2

The Problems with Low-Calorie Sweeteners

WEIGHT LOSS PARADOX

Low-calorie sweeteners have been hailed as an answer to our rising obesity problem, yet while studies are numerous, they have not provided proof that the consumption of these sweeteners is beneficial for weight loss.[1]

Obesity rates began rising in the 1980s and continue to rise, not just in America but worldwide. Over the past 30 years the number of overweight and obese people has steadily increased. It is now at epidemic proportions. To people in public health, this trend is surprising and disheartening.

Health experts had hoped that the gradual improvements in the American diet in recent years might have reversed this trend in obesity. Consumption of full-calorie soda has dropped by a quarter since the late 1990s, and there is evidence from the National Health and Nutrition Examination Survey[2] that total calorie intake has dropped for adults and children. Yet despite the reduction in sugar consumption and total calories, obesity rates keep rising.

"The trend is very unfortunate and very disappointing," says Marion Nestle, a professor in the Department of Nutrition, Food studies and Public Health at New York University. "Everybody was hoping that with the decline in sugar and soda consumption, that we'd start seeing a leveling off of adult obesity."

Much of the decrease in sugar and total calorie consumption is attributed to the use of low-calorie sweeteners. But these sweeteners have not had the positive impact they were expected to have. The idea behind sugar substitutes is that they don't contribute any appreciable amount of calories. You get the same sweet taste without the calories. If you eat the same foods, but without all of the sugar calories, your total calorie intake is lower and you lose weight—that's the theory anyway. In real life it doesn't work out that way. People switch from sugar-sweetened sodas to sugar-free versions and experience no weight loss. In fact, they tend to gain weight! It doesn't matter what type of non-caloric sweetener is used, the results are the same—weight gain, not loss.

The primary use of stevia is as a sugar substitute to aid in weight loss. High-intensity, non-caloric sweeteners such as aspartame, sucralose, and stevia are hundreds of times sweeter than sugar. Very small quantities are needed to sweeten foods to the level sugar does, providing insignificant, if any, calories.

Intuitively, people choose low-calorie sweeteners over sugar to cut down on calorie consumption in an effort to lose or maintain weight. Sugar provides a large amount of rapidly absorbable carbohydrate, leading to excessive energy intake and weight gain. Sugar has been identified as a major culprit in our obesity epidemic. Because stevia is derived from an exotic herb, it is often perceived as a "health food." Ironically, this so-called health food does not help with weight loss; in fact, it promotes weight gain, negating the primary purpose for using it.

The consumption of excess calories has been blamed for our growing obesity epidemic. The obvious solution to this problem would be to cut down on total calorie consumption. Eating less isn't easy, as it is usually accompanied by constant hunger and uncontrollable cravings that can derail the best of intentions. Food manufacturers believe they have found the solution in low- and zero-calorie sweeteners. Replacing sugar with non-caloric sweeteners allows people to indulge in their favorite foods and beverages without all the sugar calories. As long as they don't eat more than they normally would, these sugar-free foods and beverages should help prevent weight gain and aid in weight loss.

Unfortunately, just the opposite happens. Over the past two decades as non-caloric sweeteners have gained wider use and sugar consumption has declined, obesity rates have skyrocketed. In 1960, before the mass introduction of artificial sweeteners, only 14.3 percent of the American population was obese (defined as a body mass index greater than 30). Since the 1990s a dozen or more low-calorie alternative sweeteners have come onto the market— and the obesity rate has risen to an astounding 38 percent.

A billion-dollar industry has been built around low-calorie, sugar-free foods. Over the past decade there has been an explosive increase in the number of food products containing non-caloric sweeteners. Currently there are 12,000 foods on the market that contain at least one of the five FDA-approved artificial sweeteners (sucralose, aspartame, acesulfame potassium, saccharin, and neotame). Another 7,100 products contain sugar alcohols, and over 1,200 products contain stevia.[3]

It is apparent that replacing sugar with non-caloric sweeteners has not stemmed the obesity epidemic, and may actually be one of the factors that have propelled this epidemic to record-breaking heights. Doubts about the value of non-caloric sweeteners began to emerge when studies showed that those people who drank sugar-free, artificially sweetened soft drinks tended to gain more weight than those who drank the full-sugar versions.[4-5]

To help cut calories and control weight, many people choose diet sodas over the regular sugar-sweetened versions. However, researchers reported that people who drink artificially-sweetened diet soft drinks do not lose weight, but generally gain it. That's not surprising. Dr. Sharon P. Fowler and colleagues at the University of Texas Health Science Center analyzed data over an eight year period. "What was surprising was when we looked at people only drinking diet soft drinks, their risk of obesity was even higher [than those who drank the regular sodas]," she says. When the researchers took a closer look at their data, they found that nearly all of the obesity risk from drinking soft drinks came from *diet* sodas rather than the sugar-sweetened sodas. "There was a 41 percent increase in risk of being overweight for every can or bottle of diet soft drink a person consumes each day," says Fowler.[4]

It is not just overweight people, who already have a weight problem, that are affected. Normal-weight people who drink diet sodas become overweight and obese. Fowler's team looked at 1,550 subjects between the ages of 25 and 64. Of that number, 622

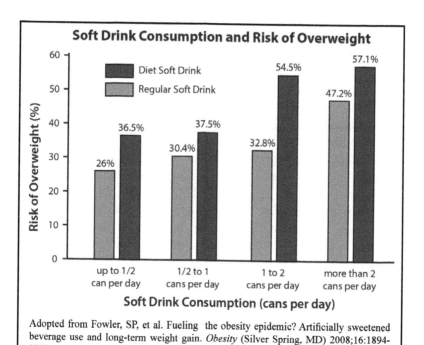

Soft Drink Consumption and Risk of Overweight

Legend: Diet Soft Drink, Regular Soft Drink

Values by consumption category:
- up to 1/2 can per day: 26%, 36.5%
- 1/2 to 1 cans per day: 30.4%, 37.5%
- 1 to 2 cans per day: 32.8%, 54.5%
- more than 2 cans per day: 47.2%, 57.1%

Y-axis: Risk of Overweight (%)
X-axis: Soft Drink Consumption (cans per day)

Adopted from Fowler, SP, et al. Fueling the obesity epidemic? Artificially sweetened beverage use and long-term weight gain. *Obesity* (Silver Spring, MD) 2008;16:1894-1900.

participants were of normal weight at the beginning of the study but after seven to eight years a third of them become overweight or obese.

Animal studies have convincingly proven that artificial sweeteners cause greater body-weight gain than sugar. When lab animals are given controlled diets containing either sugar or non-sugar sweetened foods, non-caloric sweeteners lead to greater total calorie intake, greater weight gain, and increased body fat.[6-7]

In one study, rats were separated into two groups. One group was given water sweetened with glucose, and the other was given water sweetened with saccharin. The solutions were available overnight for 10 consecutive days. The rats were also given free access to lab chow and plain water throughout the experiment. Body weight and amount of lab chow consumed were recorded for all rats on days 1, 5, and 10 of the study. The rats given the saccharin solution gained weight more rapidly and consumed a greater portion of food (see Figures 1 and 2).

The same researchers then repeated the experiment, this time using stevia. They used a brand of stevia from their local store

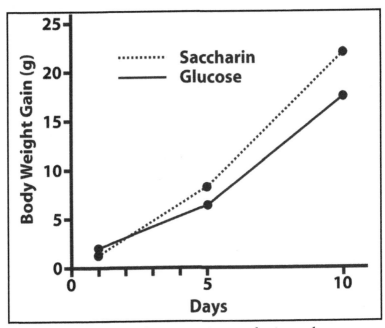

Figure 1. Weight gain after consuming saccharin or glucose.

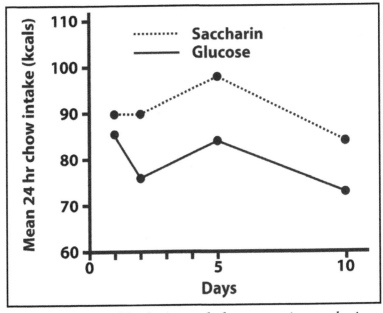

Figure 2. Amount of food consumed when consuming saccharin or glucose.

(Steviva) that was pure stevia extract (rebaudioside A) without any additives or fillers. The rats were separated into three groups. One group received a solution of stevia extract mixed in water overnight for 15 days; another group received saccharin; and the third group received a glucose solution. Lab chow and water were available throughout the experiment. Weight gain for the rats given stevia and saccharin did not differ significantly from one another at any point during the experiment. However, both the stevia and saccharin groups gained significantly more weight than the glucose-fed animals (see Figure 3). This study was conducted to determine if there was any difference in the effects of the different non-caloric sweeteners, and there wasn't. Stevia promoted weight gain just as much as saccharin did.[8]

Several large-scale human studies have also found a clear correlation between non-caloric sweetener use and weight gain.[9-11] These studies suggest that if you want to gain weight, you should be using non-caloric sweeteners in place of sugar.

Investigators from the University of Liege, Belgium, did an extensive review of published studies that included 383 studies

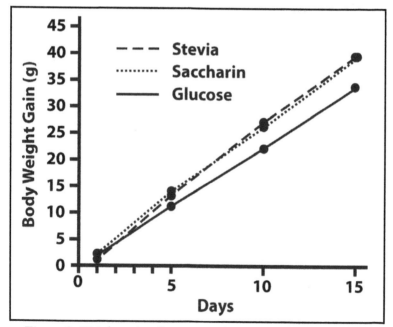

Figure 3. Weight gain after consuming stevia, saccharin, or glucose.

on the benefits and risks related to non-caloric sweeteners; 30 percent of these studies were funded by industry, 56 by non-profit organizations, and the others did not report funding sources. They found a lot of conflicting results among the studies, much of it probably due to study design and different populations. If non-caloric sweeteners truly did aid in weight loss, reduce blood sugar levels, and prevent diabetes, it should be clearly evident from an analysis of all these studies. But that is not the case.

The researchers stated: "The available studies, while numerous, do not provide proof that the consumption of artificial sweeteners as sugar substitutes is beneficial in terms of weight management, blood glucose regulation in diabetic subjects, or the incidence of type 2 diabetes."[12]

It has been calculated that consuming a non-caloric sweetener as a sugar substitute results in a decrease of daily energy intake by an average of 220 calories.[13] One pound of fat stores 3,500 calories. Thus, according to dietetic science, to lose 1 pound of fat you need to reduce your normal calorie intake by 3,500 calories. If you reduce your total calorie intake by 220 per day by substituting non-caloric sweeteners for sugar, theoretically you should lose 2 pounds every month. In one year, you should lose 24 pounds (11 kg), and in two years 48 pounds (22 kg). You should attain this rate of weight loss simply by making the change from sugar to non-caloric sweeteners, without any other adjustment to your diet or lifestyle.

But how many people who make the switch, without drastically changing their diet, see this type of weight loss? Pretty much nobody. Just switching from sugar to non-sugar sweeteners does not lead to weight loss, certainly not 24 pounds a year. Unless a person goes on a calorie-restricted diet he or she usually won't see any weight loss, and often will experience weight gain. Obviously there is something wrong here. The theory isn't working. If you reduce total calorie intake you should lose weight. But that does not happen when you use non-caloric sweeteners.

In order to make better sense of numerous studies that might have slightly different outcomes, researchers often perform a "meta-analysis," in which they combine the data from several previous studies and evaluate them together. A meta-analysis consisting of nine observational studies and 15 randomized controlled trials evaluated the effects of low-calorie sweeteners on body weight and composition. The investigators found that among the observational studies the use of non-caloric sweeteners resulted in an increase in weight, not a decrease as expected. The randomized controlled trials, where food intake was carefully controlled, showed a slight decrease in weight. The duration of the trials varied greatly, from three weeks to 18 months. During this period of time, the average drop in weight was a total of 1.7 pounds (0.8 kg).[14]

On the surface, the randomized controlled trials used in this meta-analysis appear to support the claim that non-caloric sweeteners can aid in weight loss. In fact, I'm sure some of these individual studies are used to support that claim. However, the results actually say the opposite. Based on the calculation that non-caloric sweeteners reduce daily calorie intake by 220 calories, the average loss of weight should be around 24 pounds (11 kg) per person, not a measly 1.7 pounds (0.8 kg). Even though users of non-caloric sweeteners eliminated sugar calories, their results were nearly the same as those who ate sugar-sweetened foods and consumed substantially more calories. Reducing their calorie intake using non-caloric sweeteners provided no real benefit in terms of weight loss.

While some studies do show that the use of non-caloric sweeteners results in weight loss, these very same studies demonstrate that non-caloric sweeteners cause more weight gain than sugar does. Let me explain.

Let's say we have two identical groups of people. One group uses sugar in their diet and the other group uses a non-caloric sweetener in place of the sugar. To keep things equal, let's assume both groups eat the same total number of calories. Even though calorie intake is equal, the group using the non-caloric sweetener will have a tendency to gain more weight than the sugar eating group. This is why, when people start using non-caloric sweeteners, they don't see a loss of 24 pounds in one year, or 48 pounds in two, or 72 pounds in three years, as they should.

Now let's assume that the amount of calories consumed is not equal. The group using the non-caloric sweetener consumes 220 fewer calories per day in comparison to the sugar-eating group. Using the results of the above meta-analysis, after one year they manage to lose 2 pounds more than the sugar-eating group. In effect, there is no meaningful weight loss. We know that 2 pounds of body weight equates to 7,000 calories in the diet. Even though the non-caloric sweetener group ate 80,300 fewer calories per year, weight loss only reflected a reduction of 7,000 calories per year. In other words, the non-caloric sweetener, while containing no actual calories, had an effect on the body that was equivalent to eating an extra 73,300 calories! Therefore, if you are using non-caloric sweeteners in place of sugar, you must reduce the amount you eat by about 200 calories a day to maintain the same weight as a person eating sugar-sweetened foods. Admittedly, these calculations are only theoretical, but they are based on the data from actual studies on real people.

The sum result of the randomized controlled trials in the meta-analysis showed, as did the observational studies, that non-caloric sweeteners promote more weight gain than sugar does. Those people using the non-caloric sweeteners were consuming significantly fewer calories, but they weren't losing the amount of weight corresponding to the missing calories—which reveals that the non-caloric sweeteners were somehow *promoting* weight gain independent of the calorie content of the diet.

THE SWEET TASTE

If non-caloric sweeteners reduce calorie intake, how do they cause greater weight gain? That has been a mystery; consuming fewer calories should promote weight loss, not weight gain. Researchers set out to find the answer. What they found was quite intriguing.

The answer is all about sweetness. Sugars in foods activate sweet-taste receptors on the tongue that send signals to the brain, which relays messages to the digestive tract to prepare for the corresponding sugar calories. The pancreas, in anticipation of an influx of sugar from the diet, immediately releases insulin into the bloodstream. Insulin pulls glucose out of the bloodstream, lowering blood glucose levels in anticipation of the influx of dietary sugar (glucose) that will soon enter from the meal. This way, blood sugar levels are maintained within safe and normal bounds immediately after eating.

It has been observed that lab animals consume more food when given access to water sweetened with artificial sweeteners.[15-16] Rats given saccharin solution to drink consume 10-15 percent more food than when given only water.[17] Like sugar, artificial sweeteners activate sweet-taste receptors on the tongue and send signals to the brain. But when expected calories do not come from the artificially sweetened foods, feelings of hunger intensify, which promotes overeating to compensate for the missing calories.[18-20]

A multitude of animal and human studies have now shown that non-caloric sweeteners stimulate appetite more than sugar and encourage overeating.[21-26] For example, one group of researchers fed volunteers a mid-morning snack consisting of plain unsweetened yogurt, and on another day the same yogurt sweetened to an equal sweetness with either saccharin or glucose. One hour after eating the snack they were given lunch, and the amount of food consumed was monitored. Food intake at lunchtime was significantly greater following the saccharin-sweetened yogurt compared to the plain yogurt snack consumed previously. Saccharin also stimulated further increases in calorie intake after lunch. At the end of the day, total calorie intake following the saccharin snack was greater than after the sugar-sweetened snack.[27]

The type of non-caloric sweetener doesn't seem to matter. Although researchers tested different non-caloric sweeteners, which have very different chemical properties, they all activate the sweet-taste receptors on the tongue, and increase feelings of hunger in comparison to sugar.

In another study, aspartame-sweetened water increased appetite ratings in normal-weight adult subjects, as expected; but when subjects ingested an equal dose of aspartame in capsule form, bypassing the sweet-taste receptors in the mouth, there was no effect on hunger.[28] These studies showed that the type of non-caloric sweetener made no difference. It is the activation of the sweet-taste receptors on the tongue and the lack of sugar calories, which are anticipated by the digestive system, that causes the intense increase in hunger and subsequent overeating. All non-caloric sweeteners, including stevia, will activate sweet-taste receptors and, because of their lack of calories, will heighten feelings of hunger. The sweeter the food is, the greater the effect. So, if you combine stevia with other non-caloric sweeteners, such as erythritol, to intensify the sweetness—as is often done in prepared foods—the effects are increased.

ALTERED METABOLISM

There is another, even more troublesome, effect from consuming non-caloric sweeteners. The sugar-free sweet taste not only stimulates hunger but also independently depresses metabolism, which consequently promotes weight gain![8-9, 29]

Normally, every time we eat a meal, our metabolism increases slightly for a couple of hours to help with the digestive process. This increase in metabolism burns off many of the calories consumed in the meal. Foods containing non-caloric sweeteners, however, do not trigger the normal post-meal rise in metabolism. The result is a more sluggish metabolism that causes the body to store, rather than burn, incoming calories. Many overweight people already have problems with a sluggish metabolism; they don't need to compound the problem by eating fake sweeteners. It doesn't matter what type of zero-calorie sweetener is used—saccharin, aspartame, stevia—the metabolic effects are the same. Eating foods sweetened with stevia will sabotage your weight loss efforts.

Although zero-calorie sweeteners provide no calories themselves, the effect they have on the body are the same as if they contained more calories than sugar, making them useless as weight loss aids. If weight loss is your goal, you would be better off consuming real sugar than you would stevia.

I see people all the time who struggle to lose weight on one diet or another. They follow the dietary restrictions to the letter, cut down on total calorie intake, and even exercise, but don't see the results they desire. The problem is that they eat sugar-free desserts, snacks, and beverages sweetened with non-caloric sweeteners. Because these foods are low in calories, they are often approved for those diets. This love for sweets is their downfall.

SUGAR ADDICTION

Weight loss diets are a dime a dozen. They come and they go. Few of them stick around for long. As soon as the next new weight loss fad comes around, people drop their previous diets and jump on the new one, hoping it will be more successful than the last. Over the past few decades we have seen numerous diets come and go. Some hang around longer than others. But the end results are always the same. People will lose a few pounds initially, but over time all of their weight comes right back, and they are no better off than they were before they started to diet. According to the Mayo Clinic, 95 percent of people who go on weight loss diets regain all their weight within five years. That's a 95 percent failure rate!

The more overweight a person is, the more difficult it is to lose weight and keep it off. Researchers at Kings College in London found that the chance of an obese person attaining normal body weight is 1 in 210 for men and 1 in 124 for women, increasing to 1 in 1,290 for men and 1 in 677 for women with severe obesity.[30] In other words, if you are obese it is nearly impossible for you to attain permanent weight loss.

Why is it so difficult to lose weight and keep it off? The answer is addiction. Addiction to sugar is the universal factor that dooms all weight loss diets to failure. Nearly all overweight people are addicted to sugar and sweets; that's why they are overweight. Everybody loves sweet-tasting foods. The sweet taste triggers pleasure centers in our brains. In some people this stimulus becomes addictive and uncontrollable. Overeating and weight gain is the result.

Most weight loss diets ignore sugar addiction. Some give it lip service but then provide a variety of "low-calorie" dessert recipes to appease the dieter and fuel the addiction. In fact, the majority of weight loss diets—whether they are low-fat, low-carb, high-fat, high-protein, vegan, or whatever—purposely fuel this addiction in order to sell books, diet programs, and food products. For example, look at all of the desserts and sweet foods sold by companies such as Weight Watchers, Jenny Craig, and Atkins. Their offerings are a sugar addict's dream. Even so-called "low-carb" diets can be loaded with sweets: you can buy low-carb candy bars (protein bars), cakes, cookies, and other desserts. There are plenty to choose from. Even though these desserts are not sweetened with sugar, they are sweetened with non-caloric sweeteners, which are just as addicting. Diet cookbooks, even low-carb and ketogenic

cookbooks, that profess abstinence from sugar and high-carb foods are loaded with dessert recipes to entice customers to buy the books. There are even entire books devoted to low-carb and ketogenic desserts! These recipes are all shams that keep sugar cravings alive and guarantee your eventual failure.

Diets rich in sugar and other sweeteners drive our current obesity epidemic. The biggest stumbling block to successful weight loss is sugar addiction. No weight loss diet will ever be successful as long as this addiction is maintained.

Overconsumption of sugar-dense foods and beverages is initially motivated by the pleasure of sweet taste, and is often compared to drug addiction. This may sound like an overstatement, but sugar addiction can cause persistent compulsive behavior, severe anxiety, loss of sound judgment, and even physical symptoms, dooming any weight loss diet to failure.

A number of studies have shown that sugar and non-caloric sweeteners can be more addictive than cocaine—one of the most addictive and harmful substances currently known.[31-33]

For example, a team of researchers from France and Australia gave lab rats free access to both cocaine and sugar-sweetened water. When exposed to both substances, their preference for sugar was far greater than the desire for cocaine. Even rats who were already addicted to cocaine quickly switched their addiction to sugar when offered the choice; they were also more willing to work for sugar than for cocaine. This study demonstrated that sugar has a stronger addictive tendency than cocaine.

One question the researchers wanted to answer was, does the preference for sugar result from the chemical properties of sugar, or from its sweet taste? To answer this, the researchers also tested the rats using the artificial sweetener saccharin, which is completely different chemically from sugar. The results were the same. The type of sweetener didn't matter: it was the sweet taste that triggered the powerful effect.[31]

Other studies have shown that, when given the choice between plain water and water sweetened with sugar or saccharin, mice would gorge themselves with the sweetened water, drinking three times as much as they normally would, clearly displaying addictive behavior. Their conclusion was that sweetened foods and beverages, regardless of the sweetener used, generates a stimulus that is so strong it can override normal behavior and self-control, and thus lead to addiction.

Stevia appears to have a similar effect. Researchers gave rats and mice the choice between water sweetened with saccharin or rebaudioside A (stevia), and unsweetened water. The animal's preference for rebaudioside A was just as strong as it was for saccharin. In a direct comparison between saccharin and rebaudioside A, however, the former was preferred.[34] Interestingly, rebaudioside A was preferred over other artificial sweeteners such as aspartame and cyclamate.

The sweet taste itself triggers physiological responses in the body and the brain that are in many ways very similar to addictive drugs. For instance, both sweet taste and drugs of abuse stimulate dopamine signaling in areas of the brain involved in reward-processing and learning.[35-36] Also, there is a cross-tolerance and a cross-dependence between sugar and addictive drugs; for example, animals with a long history of sugar consumption actually became tolerant of (desensitized to) the analgesic effects of morphine.[37]

Additionally, studies have found that the overconsumption of sweetened foods, when stopped, can lead to withdrawal symptoms similar to cocaine addiction.[38] Finally, MRI studies in humans have discovered adaptations in the brains of obese individuals that mimic those observed in cocaine addicts.[39-40]

It is not so much a *sugar* addition as it is a *sweet* addiction, as any type of sweetener can cause and maintain addiction. Whether the sweet taste comes from sugar, aspartame, xylitol, or stevia, the results are the same. The sweet taste perceived by the receptors on the tongue signals the brain, initiating cravings and addictions. The receptors do not distinguish any difference between sugar and stevia, the same signals are relayed to the brain.

Zero- or low-calorie sweeteners do not help with weight loss or with overcoming sugar addiction. If you are trying to lose weight, sugar substitutes are your enemies, not your friends. Sugar substitutes give you a false sense of security while fueling the fire of sugar addition. Using sugar substitutes, including stevia, keeps sugar addictions and cravings and bad habits alive.

Many people use sugar substitutes as a means to breaking their addictions to sugar—but end up addicted to these other sweeteners while maintaining their sugar addiction.

I became aware of stevia addiction before I read the above studies. I've seen people overuse it and abuse it, and even develop a tolerance to its bitter aftertaste. One such person gave me a sample of the stevia-sweetened water she was drinking, and it almost knocked me over with its intense sweet-bitter taste. I've eaten plenty of foods and drinks sweetened with stevia, but this was way overboard. I couldn't imagine anyone enjoying it—yet this person drank stevia-sweetened water all day long. She also had a very difficult time losing weight, even though she ate very little and followed a strict low-carb diet.

People have fallen in love with stevia because of their addiction to sugar. They are looking for a natural substitute for sugar that will give them all the sweet satisfaction of sugar without the detrimental side effects—it's not going to happen! The only way to kick the addiction to sweets is to abstain from them.

If you use more than just a tiny amount of stevia you will notice clearly the characteristically bitter aftertaste. If you do *not* notice the bitter aftertaste, the brand you are using is heavily adulterated with other sweeteners, or you have built up a tolerance to this effect, which is a clear sign of stevia addiction.

People so desperately want to believe that there is a natural, harmless sugar substitute that they overlook the obvious and close their eyes to the evidence. Some people get irate if you say anything bad about stevia; they almost get violent in its defense. It reminds me of an alcohol or drug addict who refuses to admit his addiction and is trying to defend his actions. Stevia (as purified steviol glycosides) is a drug, an addictive drug, so it stands to reason that some people will become so incensed about any challenge to its safety and its effectiveness as a sugar substitute. Getting angry is a sign of addiction. Addiction overrides common sense and reason, and motivates a person to make unwise eating choices.

And of course, there are those who make money promoting and selling stevia. They will be greatly disappointed by these revelations about stevia, may even refuse to acknowledge them, and may put up a bitter fight in its defense. When money gets involved people can get highly emotional.

ANTI-KETOGENIC

A ketogenic diet is one that is very low in carbohydrate, high in fat, with moderate protein. With a ketogenic diet, the body gets most of its energy from burning fat rather than glucose. The fat that is used to power the body's cells comes from both stored body fat and dietary fat. Since stored fat is utilized to satisfy energy needs, excess body fat is readily reduced, making a ketogenic diet ideal for weight loss when total calories are also restricted.

The diet was originally developed in the 1920s to treat epilepsy, which it did very successfully. It is still the most successful treatment for epilepsy, reducing seizures significantly, and in many cases bringing about a complete and permanent cure. The success with epilepsy has led to its use in treating other neurological conditions, such as Alzheimer's, Parkinson's, ALS, stroke, autism, developmental disorders, and depression. Before the discovery of insulin in the 1920s, a ketogenic diet was used to treat type 1 diabetes. Today it is the most effective treatment for both type 1 and type 2 diabetes, and has proven invaluable in the treatment of metabolic syndrome, which is characterized by abdominal obesity, elevated blood pressure, insulin resistance, and abnormal cholesterol levels. In recent years, it has become popular as an effective means for weight loss. It has also been recommended for the treatment of heart disease, cancer, glaucoma, macular degeneration, alcoholism, and substance abuse, and for enhancing athletic performance and endurance.[41-44] The ketogenic diet exerts a remarkable therapeutic effect that improves overall health.

Normally, we depend on glucose to supply most of our energy needs. We get glucose primarily from the carbohydrates in our diet. Between meals and when we are asleep, fasting, or not eating carbohydrates, blood glucose levels fall and the body begins mobilizing stored fat to supply its energy needs. Some of this fat is converted into a high-density type of fuel called ketones. Ketones are a highly efficient form of fuel that delivers more energy than glucose or fat. A person is said to be in ketosis when they are burning more fat and ketones for energy than they are glucose. When someone goes on a ketogenic diet, he or she goes into ketosis, often referred to as nutritional ketosis because it is diet-induced.

In order to keep blood glucose levels low enough to get into ketosis, the diet must be very low in carbohydrate. Carbohydrate consumption must be limited to less than 50 grams per day, and generally is much lower. Most people ordinarily consume 5 to 6 times this amount. The less carbohydrate consumed, the higher blood ketone level rises and presumably the greater the therapeutic effect.

Blood ketone levels can be measured with ketone test strips or a blood monitor. Many people use these tools to monitor their level of ketosis and to keep their diet on track. If they eat too much carbohydrate, even if only by a few grams, they can see it. The test strips, which are the cheapest and most convenient to use, give a general indication of ketone level designated by "none," "trace," "low," "medium," and "high."

Because carbohydrate consumption must be kept very low, all sugar and sweets are eliminated in the diet. If any sweetening is desired, stevia is often recommended. The problem is that stevia is anti-ketogenic. It can knock a person out of ketosis in a heartbeat. In fact, stevia is even more anti-ketogenic than pure sugar.

Lisa had been on a strict ketogenic diet, limiting her carb intake to less than 25 grams daily. She had been following this way of eating for several weeks, but her ketone levels, as measured by ketosis test strips, were only in the "trace" to "small" range. Her husband, who was eating the same foods as she was, had a "high" ketone reading. The only difference in their food consumption was that Lisa routinely drank water sweetened with stevia. Before starting the new diet she was addicted to sugar, and was controlling the addiction, or so she thought, by drinking stevia-sweetened water. Suspecting stevia to be the problem, she discontinued using it and her ketone levels abruptly rose.

I've seen this time and time again: people following a ketogenic diet eating so-called ketogenic-friendly desserts sweetened with stevia, and wondering why they can't get into ketosis (or to moderate to high ketosis) and why they aren't losing weight.

Other non-caloric sweeteners aren't any better. Sugar alcohols are just as bad. Chewing gum, toothpaste, and mouthwash sweetened with xylitol can kick you out of ketosis. Toothpaste and mouthwash aren't even swallowed; the sweet taste alone is enough to affect metabolism to a degree that it arrests ketone production.

I've seen a greater drop in ketosis in people who ate a low-carb food sweetened with just a few drops of stevia, than in people who ate a slice of high-carb cake made with white flour and processed sugar. Just a tiny amount of stevia can affect metabolic processes so dramatically that it immediately turns off ketone production. If you are on a ketogenic diet for the purpose of losing weight, controlling diabetes, or treating neurological disorders like Alzheimer's or epilepsy, the therapeutic effect is stopped or at least greatly reduced when stevia is consumed. If it can do this, then it may have other undesirable effects as well.

We always have some level of ketones in our bloodstream. They are highest when we don't eat for a period of time (fasting) or when we don't eat carbohydrates (ketogenic dieting). Ketones are essential for good brain health. When blood ketone levels rise, they trigger the activation of special proteins in the brain called brain derived neurotrophic factors (BDNF), which regulate healing and repair and ease inflammation (a common problem with many brain disorders). BDNF also stimulate the growth and development of new brain cells, keeping the brain healthy and functional.

Without ketones activating BDNF on a regular basis, brain aging and degeneration accelerates. Stevia blocks ketone production, preventing the formation of BDNF and their therapeutic effect. Therefore, stevia as well as other fake sweeteners contribute to or increase the risk of memory loss, stroke, Alzheimer's, and other neurological disorders.

Why is stevia anti-ketogenic? I searched the medical literature for an answer, and found an important clue in a study published in the journal *Research Communications in Chemical Pathology and Pharmacology*. Our bodies produce ketones from fat. When blood sugar levels are low, the body releases fatty acids from fat cells. Our cells use fatty acids like they do glucose, to produce energy. The liver converts some of these fatty acids into ketones—the preferred fuel for the brain and heart. During a fast, ketones serve as a major source of energy for the body.

The objective of the study was to observe the influence stevia has on liver glycogen levels when in a state of ketosis. Glucose is stored in the form of glycogen (a type of starch) in the liver and is released as needed to keep blood sugar levels in balance throughout the day. In the study, investigators put rats on a water-only fast for 24 and 48 hours, which put them into a state of ketosis. The investigators then fed the rats water sweetened with stevia extract (stevioside), without any other type of food. The stevia feeding stimulated glycogen synthesis and storage in the livers of the rats.[45] This is an important observation, because when the liver starts to produce and store glycogen, it stops producing ketones. Thus, we have the reason why stevia lowers blood ketone levels.

The liver converts sugars and carbohydrates into glucose—blood sugar. It also produces ketones, but not at the same time. It produces one or the other depending on how much sugar and carbohydrate, and to some extent protein, is in the diet. Approximately 50 percent of the protein in our diet can, if needed, be converted into glucose. Normally this doesn't happen, because we eat plenty of carbohydrate-rich food to supply all the glucose we need on a daily basis. When there are carbohydrates or a lot of protein in the diet, ketone production is turned off.

Stevia activates the production of glycogen because of its sweet taste. When the taste receptors sense the sweetness of stevia, a signal is sent to the pancreas to release insulin. Insulin is a storage hormone. It triggers the transformation of glucose into fat, and stores it in the fat cells; it also stimulates the conversion of glucose into glycogen, and stores that in the liver. Consequently, ketone production is turned off. Ketosis is the opposite process; the liver takes fat out of storage to produce ketones.

Apparently, when carbohydrate is not consumed, as during a fast or a ketogenic diet, eating stevia activates metabolic processes that can break down protein into glucose for storage in the liver. If there is not enough dietary carbohydrate or protein, then muscle and organ tissue can be cannibalized to supply it. In the study above, because the rats were not eating anything, the glucose must have come from the breakdown of muscle tissue. Therefore, if you are on a low-carb or ketogenic diet and eat foods containing stevia, it will cause muscle tissues to be broken down and converted into glycogen.

Sugar alcohols, such as xylitol, also block ketone production. It is likely owing to the same cause. Stevia and xylitol are chemically very different but have a similar effect, so apparently it is the sweet taste that is triggering glycogen storage. Chewing xylitol-sweetened gum or brushing your teeth with xylitol-sweetened toothpaste stops ketone production even though xylitol is not actually consumed. As mentioned, the sweet taste alone is enough to trigger the reaction, which means that *any* low-calorie sweetener would trigger a similar response. This would explain why many people have reported that using aspartame and other artificial sweeteners kicks them out of ketosis.

This can be a very serious issue. If people go on a ketogenic diet for therapeutic reasons—such as to treat obesity, manage diabetes, reduce insulin resistance, reverse metabolic syndrome, reverse neurological disorders (Alzheimer's, Parkinson's, epilepsy, autism, etc.), fight cancer, or to achieve any of the documented benefits associated with the ketogenic diet—they *must* eat foods that are truly ketogenic. Eating foods sweetened with stevia and other low-calorie sweeteners will prevent ketosis and its associated therapeutic effects. A meal that may otherwise be ketogenic can become anti-ketogenic by the addition of stevia and therefore, do more harm than good.

I have not found a ketogenic cookbook I could recommend to people because they all contain recipes using low-calorie sweeteners that make the foods anti-ketogenic. For this reason, I wrote a ketogenic cookbook that contains no sweeteners except for a limited amount of fresh fruit. The book is titled *Dr. Fife's Keto Cookery: Nutritious and Delicious Ketogenic Recipes for Healthy Living.*

3

Health Claims

THE WONDER HERB

The Internet is awash with fantastic health claims about stevia. Websites paint stevia as a wonder product useful for treating obesity, diabetes, hypertension, infections, dental cavities, liver disease, digestive problems, heart disease, sugar cravings, and much more.

Some of the health claims are based on folklore, hearsay, and a bit of wishful thinking; others are based on hypothetical extrapolations from preliminary research, all of which lack solid scientific proof. Only a few of the purported health benefits are supported by a reasonable amount of documented evidence. By far the most recognized effect of stevia is its intense sweet taste, which has led to the assumption that it must be useful as an aid in weight loss when used in place of sugar—an assumption we have seen to be false.

Stevia exhibits a hypoglycemic effect (lowers blood sugar) and appears to stimulate insulin release, and some claim it also improves insulin sensitivity and secretion, all of which suggests its possible value in treating diabetes; thus its general recommendation for this purpose.

It is reported to lower elevated blood pressure, which is a major risk factor for heart disease; thus the claim it protects the heart.

Some researchers have suggested that stevia may possess anti-inflammatory and antioxidant properties; these reports are based on animal and tissue studies in the laboratory.

Certain oral bacteria feed on sugar and promote cavities. Since stevia is not broken down by oral bacteria, using it in place of sugar in the diet should theoretically reduce the risk of cavities. Stevia extract has also been shown to have antimicrobial action against certain bacteria, viruses, and fungi, including the *Streptococcus mutans* bacterium that is the major cause of dental cavities. These aspects provide the bases for the claim that stevia can prevent cavities.

In lab animals, stevia has been shown to ameliorate the harmful effects of certain toxins that adversely affect the liver and kidneys, suggesting a possible benefit to liver and kidney health.

Is stevia a wonder herb with miraculous healing properties? Or is it the product of over-zealous marketing? Let's take a closer look.

EFFECTS FROM NON-SWEET COMPONENTS

There is a lot of confusion and misunderstanding about the health benefits attributed to stevia, because studies often use the term "stevia extract" to describe different stevia-based products, each containing different compounds. Salespeople, marketers, and product labels also use this term, implying the product you purchase in the store is the same as that used in studies, which may not be the case.

Many of these studies use crude hot-water extracts of the leaf and stem, not the purified rebaudioside A you purchase in the store. Hot-water extracts contain a multitude of chemicals found in the plant; any one or more could bring about the reported effects. Most of the studies reporting the anti-inflammatory, antioxidant, antimicrobial, and protective properties on the liver and kidneys were done using leaf and stem extracts, not purified stevioside or rebaudioside A. These beneficial effects were attributed to other compounds in the plant, such as carotenoids, flavonoids, polyphenols, and anionic polysaccharides, which are not found in stevia sweeteners.[1-4]

Stevia products: fresh and dried leaf, liquid extract, tablets, and powder.

For example, the toxin streptozotocin is a potent oxidizing agent that can cause damage to multiple organs, notably the pancreas, kidneys, and liver. Researchers use streptozotocin to purposely damage the insulin-producing cells in the pancreas of lab animals to induce diabetes. In one study, supplementing the diet given to lab rats with crude stevia extract reduced injury from streptozotocin. The authors gave the credit to the antioxidant polyphenols in the extract. Their conclusion was that stevia leaf extract could potentially protect rats against streptozotocin-induced diabetes, reduce the risks of oxidative stress, and ameliorate liver and kidney damage.[5] Stevia sweeteners do not contain polyphenols, so this study has no relevance to the stevia sweeteners sold in the store. But that has not stopped stevia promoters from pointing to this study as evidence that it is beneficial to diabetics and can help protect the liver and kidneys. Most of the studies that show health benefits of stevia are due to other compounds in the whole leaf and have no association with steviol glycosides used in sweeteners. Realistically, eating stevia-sweetened foods will not reverse diabetes, prevent tooth decay, ward off infections, or protect the pancreas, liver, or kidneys from harmful toxins.

LOWER BLOOD PRESSURE

A number of studies have shown that purified stevia extract can decrease blood pressure. This sounds impressive. Hypertension or high blood pressure puts a tremendous strain on the heart and is one of the major risk factors for heart disease. The effects of stevia are so minimal, however, that it would require huge doses to provide even a small benefit on blood pressure.

Researchers at Taipei Medical University in Taiwan investigated the long-term (two-year) efficacy of steviol glycoside extract in patients with mild hypertension.[6] The study's subjects took capsules containing 500 mg of steviol glycosides three times daily. After two years they showed a mean decrease in blood pressure of a mere 5 percent. The subjects were taking a total of 1,500 mg of concentrated stevia extract every day. This would be equivalent to drinking 8 cans (12 ounces/340 g each) of stevia-sweetened beverage every day—far more than most people would consume. Consequently, drinking one to three cans daily, which would be more realistic, would have little beneficial effect on blood pressure, a result verified by subsequent research.[7]

Stevia does not lower blood pressure by correcting the underlying problem. The reason for the high blood pressure remains. All it does is mask the problem. The steviol glycosides in stevia have a diuretic effect that increases urine output, thus lowering blood pressure. The same effect can be obtained from any number of other natural and drug-based diuretics.

LOWER BLOOD SUGAR

Next to its distinctive sweet taste, the most pronounced effect stevia has is on insulin release and blood sugar levels. Interestingly, it is the sweet taste that controls these effects. Crude stevia leaf extract and purified steviol glycosides taken orally have shown to reduce blood glucose levels in both diabetic and non-diabetic human subjects.[8-9] The drop in blood glucose is the body's response to insulin release, triggered by the sweet taste of stevia.

Normally, when you eat a meal, the sugars and carbohydrates in the food are converted into glucose and sent into the bloodstream. Consequently, immediately after a meal, especially a high-carb meal, blood glucose levels rise. The body tries very hard to maintain blood glucose levels within a narrow range. If blood glucose levels go too far one way or the other, it can lead to serious health problems, including coma and death.

When a meal that contains sugar is eaten, the pancreas instantly begins to release insulin in anticipation of receiving a high influx of glucose from the sugar into the bloodstream. This surge of insulin immediately begins to pull glucose out of the bloodstream, and glucose levels begin to drop. As the surge of glucose from the meal enters the bloodstream, blood glucose is quickly restored. This way, blood sugar can be held within the bounds that are safe, and prevents dangerously high glucose spikes immediately after eating.

This reaction is observed in both animals and humans. Our taste buds detect five different taste sensations—sweet, sour, salty, bitter, and umami (savory)—but only the sweet taste triggers the release of insulin.[10]

It is the sweet taste itself, and not the particular type of food, that triggers the release of insulin; and the sweeter the taste, the greater the insulin release and the lower the blood glucose drops. When a non-caloric sweetener is consumed, the brain reacts to it just as if it were sugar, and insulin is released. The influx of insulin into the bloodstream pulls glucose out, lowering blood sugar. Since the low-sugar meal containing stevia does not replenish the glucose that was removed, blood sugar levels remain depressed for a few hours afterwards. Lowered blood sugar stimulates hunger and encourages cravings and overeating.

The side effects of eating non-caloric sweeteners include an initial release of insulin, lowered blood glucose levels, and hunger, which encourages overeating. The sweet taste is the trigger, regardless of the specific sweetener. All of these effects are clearly demonstrated in both human and animal feeding studies using a variety of non-caloric sweeteners, including stevia. The oral administration of stevia extracts have been shown to reduce blood glucose levels in a dose-dependent manner and stimulate insulin secretion just as other non-caloric sweeteners do.[11]

Yet some researchers claim that stevioside itself may improve insulin secretion and insulin sensitivity, and therefore may be useful for diabetics.[12] But other studies have shown that stevioside or rebaudioside A, consumed in capsule form, which bypasses the sweet taste, has no insulin-stimulating or glucose-lowering effect.[13]

For example, researchers carried out a randomized, double-blind, placebo-controlled investigation of the effect of daily consumption of 1,000 mg of rebaudioside A on the blood glucose in type 2 diabetic individuals. While maintaining a stable diet, subjects received four 250 mg capsules (two with morning meal and two with evening meal) per day of rebaudioside A or a placebo for 16 weeks. There was no significant difference in blood glucose levels between the treatment and placebo groups.[14]

Another group of investigators performed a similar study examining the effect of stevioside on blood glucose taken in capsule form. This randomized, double-blind, placebo-controlled, long-term study examined three groups of subjects: type 1 diabetics, type 2 diabetics, and non-diabetics.[15] The results were the same: no effect.

To conclude: these studies demonstrated that the physiological effects attributed to steviol glycosides were not caused by the chemicals themselves, but by their sweet taste.

4

Safety Concerns

THE FDA BAN

Stevia has a controversial safety history. Whole leaf stevia has been used in Brazil and Paraguay for centuries without any apparent problems. The Japanese have used it for decades without reporting any significant harm. With the growing interest in alternative sweeteners, stevia has been seen as a harmless and even healthful alternative.

Surprisingly, in 1991 the Food and Drug Administration (FDA) of the United States labeled stevia an "unsafe food additive," restricted its import, and banned its use as a food additive. Many people viewed stevia as a harmless herb and couldn't understand the reason for such a ban. The FDA's position on stevia was based on a growing number of reports in the medical literature that raised concerns about its safety. Animal studies linked stevia with adverse effects on blood pressure, blood sugar, kidney and heart function, and fertility and reproductive development; concerns were also raised about possible genetic mutations. These concerns have been based on studies dating back as far as the 1960s and continuing up through the 2000s, which is one of the reasons the FDA still has not lifted the ban on stevia leaf as a food ingredient.

The FDA wasn't alone in banning stevia. The concerns raised in the medical literature prompted the European Commission in 1999 to ban stevia's use in foods and beverages in the European

Union, pending further research. Canada, Australia, New Zealand, and other countries followed suit and banned stevia and its extracts for use as food additives.

REPRODUCTIVE HEALTH

These bans weren't haphazard decisions, but were based on published medical research at the time. Concern about reproductive health was prompted by traditional uses for stevia among the Paraguayan Indian tribes, who use it as an oral contraceptive.[1] For this reason, some of the earliest research on stevia involved its possible effects on reproduction and fertility.[2] These studies raised a red flag. For instance, researchers observed reduced fertility in female rats exposed to approximately 0.5 gram of dried plant material per day through drinking water prior to and throughout mating. In male rats, exposure to 1.33 grams of dried leaves up to twice a day for 60 days was associated with lower testosterone levels and lower testis, epididymis and seminal vesicle weights, and lower sperm counts.[3-4]

These studies are not conclusive. Other studies have not shown any harmful effects on reproduction. However, if someone is trying to conceive, or is having a difficult time conceiving, it probably would be a good idea to play it safe and avoid stevia.

Stevia may not be the only non-caloric sweetener that might adversely affect the growth and development of an unborn child. In one study the daily intake of artificially sweetened soft drinks was shown to increase the risk of preterm delivery. This is significant because a premature infant is at far greater risk of experiencing health problems both in infancy and later in life. The risk increases as artificially sweetened beverage usage increases.[5]

It seems like a good idea not to consume any non-caloric sweetener during pregnancy.

BLOOD SUGAR

Stevia's well-known effect of lowering blood sugar is often touted as a benefit, specifically for type 2 diabetics. But this effect may actually be harmful to them. A characteristic of diabetes is chronically elevated blood glucose levels, and a simple means for lowering these levels is generally viewed as beneficial, even if it is only temporary. As mentioned earlier, the way stevia lowers blood sugar is by triggering a rapid release of insulin into the bloodstream. For a type 2 diabetic, this isn't necessarily a good thing. Type 2 diabetics already produce excessive amounts of insulin. In type 2 diabetes the cells in the body have become insulin resistant, meaning the cells do not respond as well as they should to the action of insulin, and therefore the pancreas must produce a much greater amount of insulin to control blood glucose levels. This puts a heavy strain on the pancreas. After years of excessive stress from producing large amounts of insulin, the insulin-producing cells in the pancreas begin to burn out and die. For this reason, many type 2 diabetics become insulin–dependent, and must have insulin injections on a regular basis to supplement the diminishing amount secreted by a failing pancreas. Consuming stevia-sweetened foods forces the pancreas to work harder, increasing the risk of having them burn out quicker.

At least 90 percent of diabetics are type 2. Most of the remaining 10 percent are type 1. In type 1 diabetics the situation with stevia is different. Their pancreases don't produce insulin at all, so they require daily injections to keep their blood sugar levels down. Stevia would have no blood sugar–lowering effect on them because they cannot produce insulin in response to its sweet taste.

LIVER FUNCTION

Some studies suggest that stevia may cause liver damage. For instance, the patients in one study were given 1,000 mg/day of rebaudioside A (the equivalent of drinking eight 8-ounce servings of stevia sweetened beverage) for 16 weeks.[6] The investigators reported no adverse effects on blood sugar or blood pressure compared to a placebo, suggesting the safety of rebaudioside A for diabetics. However, there was a small but *significant* increase in alanine transaminase (ALT) levels in the rebaudioside A group compared to the placebo group. ALT is an enzyme found mainly in the liver, but also in smaller amounts in the kidneys, heart, and pancreas. Doctors measure ALT levels in the blood to check for liver damage. The blood normally contains very little ALT, but when the liver is damaged or diseased, it releases ALT into the bloodstream, increasing ALT levels. ALT tests are performed by doctors to identify liver disease, especially cirrhosis and hepatitis caused by alcohol, drugs, or viruses, and to track the effects of medicines that can damage the liver.

Although the elevation in ALT levels is an important indication of something happening to the liver, and possibly other organs, the researchers in this study brushed the finding aside, suggesting (without any valid reason) that it was due to random variation, and claiming it had no clinical significance since mean levels of ALT in the subjects stayed within normal range. No explanation of "clinically significant" or "normal range" was provided by the investigators. This study only lasted 16 weeks, so long-term effects are unknown. The fact that a standard marker for liver damage was present is a cause for some concern and needs further investigation.

Other studies have shown large, but inconsistent, reductions in bile acids.[7-8] The liver produces bile, which is necessary for the emulsification and digestion of dietary fats. Rebaudioside A, stevioside and other steviol glycosides travel down the small intestine unchanged. In the colon they are reduced by gut bacteria into steviol, which is absorbed into the bloodstream. Steviol provides no useful function, and must be cleared out of the bloodstream like any other foreign body. It is believed that the removal and excretion of steviol by the liver causes the changes in bile metabolism identified in these studies. Obviously stevia puts some degree of stress on the liver, resulting in elevated ALT levels and altered bile output.

MUTAGENICITY

Another potentially serious concern is the possibly of genetic toxicity—the ability of a substance to damage the genetic information within a cell causing mutations, which could lead to deformities or cancer. Some studies suggest that stevia can induce gene mutations, chromosome alterations, and DNA damage.[9-13]

Stevioside (88.62% purity) was reported to be positive in an *in vivo* test of DNA double-strand breaks in cells of peripheral blood, spleen, liver, and brain of rats following administration at 4 mg/ml in drinking water for 45 days; DNA damage was seen after only five weeks of exposure.[14] Japanese researchers reported that stevioside forms a epoxide metabolite (steviol-16alpha,17) when incubated for 48 hours in intestinal microflora. Epoxides are of concern because they are highly reactive with DNA.[15] This may not be a concern for those who have a good working digestive system that passes food though the body in less than 36 hours or so. But for those people whose digestive systems work a little slower, or who are frequently constipated, it allows more time for stevia epoxides to form.

In a 2008 review of published studies on stevia by the UCLA Department of Environmental Health Sciences and Molecular Toxicology, researchers identified eight studies that showed genotoxic activity for rebaudioside A and stevioside.[16] However, another 23 studies failed to find any significant genotoxic effect. Study results can vary depending on many factors such as different product formulations, diverse experimental designs, different subject populations and cohort sizes, and varying durations of treatment and the results could all be valid under the conditions used. Although the majority of studies failed to find a genotoxic effect, the fact remains that eight independent research teams did find a connection, which means that this issue is not clearly decided. The authors of the review stated that there were still too many unanswered questions and concerns about stevia and its glycosides at that point, and that it should not be granted GRAS (Generally Regarded As Safe) status, which the FDA was considering at the time.

There has been some concern that if stevia is mutagenic, it may also be carcinogenic. However, no evidence has been reported that rebaudioside A or stevioside are carcinogenic.[17-18]

Some studies even suggest that steviol glycosides are anti-carcinogenic and might actually protect against cancer.[19-20] More research needs to be done to verify this possibility.

While rebaudioside A and stevioside appear not to be carcinogenic or to have any mutagenic effect, the same may be not true with steviol, the metabolic end product of these glycosides. Some studies suggest that steviol is mutagenic.[9, 21]

In some studies rebaudioside A and stevioside appear to demonstrate some antimicrobial activity. However, it has been suggested that the reason for this is due to the mutagenic effect of steviol. Studies have shown steviol to cause genetic mutations in some microorganisms, which could greatly impair their ability to reproduce and thrive.[9, 11, 22-23] The problem here is that steviol may be mutagenic not only to microorganisms, but to our cells as well. Also, if it is mutagenic to bacteria, what effect does it have on the organisms living in our digestive tract? It could seriously alter the gut microbiome.

KIDNEY FUNCTION

Kidney function is another potentially serious concern. Steviol, the metabolite of steviol glycosides, is a toxin that is absorbed into the bloodstream and is removed primarily in the urine. To facilitate its rapid removal, the body steps up urine excretion. As a result, chronic consumption of stevia may lead to dieresis (fluid loss/dehydration) and natriuresis (excessive excretion of sodium).[24-25]

Safety studies published in 2008 by researchers from Cargill noted that high levels of rebaudioside A, from 12 to 14g/kg body weight/day administered to rats for four weeks, not only affected bile levels, but showed higher mean creatinine and urea levels.[7] There were also some differences between treated and control animals, respectively, in slightly lower epididymis weight and lower relative heart, kidney, and adrenal weights. In males, testes weight was slightly lower in treated than in control animals. The epididymis is part of the male reproductive system and is attached to testes, and functions as a sperm storage reservoir. Lower epididymis and testes weight suggests some adverse effect in the reproductive system, as noted in other studies.

The increase in creatinine levels is of significant interest. Elevated creatinine is a sign of kidney damage or kidney disease. Creatinine is a waste product that is generated from energy production in muscles. Creatinine is transported through the bloodstream to the kidneys where it is filtered out and disposed of in the urine. Since muscle mass in the body is relatively constant from day to day, creatinine production and elimination remains relatively unchanged on a daily basis. Measuring blood creatinine levels has become a reliable indicator of kidney function. As the kidneys become impaired for any reason, creatinine levels rise due to poor clearance by the kidneys. An abnormally high level of creatinine is a warning sign of possible kidney damage. Any rise in creatinine, no matter how small, is an indication of kidney stress and lowered efficiency.

The Cargill researchers stated that these changes were of little concern because they were all within normal limits, and that an examination of the tissues did not reveal any obvious damage. Since the amount of rebaudioside A given to the test animals was proportionately far larger than what a person would normally consume in a day, the researchers' conclusion was that rebaudioside A posed little danger. The problem here is that rebaudioside A did have an adverse effect on kidney and reproductive health, which should have served as a warning sign of possible problems. This study only lasted four weeks; the long-term effects are unknown. While smaller doses per day may not show any significant changes, daily use for many months or years could possibly have significant consequences.

A study published by Japanese researchers a few years earlier also brought attention to stevia's potential effect on kidney health. The researchers gave a single oral dose of stevia to the test animals and observed them for 14 days. Stevia caused death in some of the rats and mice, but the effect was especially pronounced in hamsters. In hamsters, most deaths occurred within 48 hours after dosing. Signs of toxicity included drowsiness, weakness, increased feces, decreased activity, and lethargy, all of which were noted prior to death. Examination of the organs revealed congestion and severe degeneration of the liver and kidneys. Acute kidney failure was offered by the authors as the possible cause of death.[26]

The dose given to the animals was 15g/kg body weight, which far exceeds the dose that is generally considered the practical upper limit for rodents. This study didn't fairly evaluate the toxicity of a reasonable dose, but it did demonstrate that ingesting too much stevia can be fatally toxic.

PROTEIN BREAKDOWN

The body's reaction to the sweet taste of stevia causes glucose to be pulled out of the blood, converted into glycogen, and stored in the liver. This leads to a drop in blood glucose. Low blood sugar is a serious matter. When the anticipated sugar calories don't come to boost blood sugar back to normal, the body becomes frantic, and epinephrine and cortisol surge to mobilize glucose from other sources.

If the diet does not supply enough carbohydrate to replenish blood sugar levels, the body must make its own glucose from protein. It gets this protein by cannibalizing lean tissues, such as muscles. Many of the amino acids that make up protein can be broken down and converted into glucose. However, protein isn't a particularly good source for glucose, so a lot of it must be broken down to get what is needed.

Stevia is not going to affect everyone's blood sugar in the same way or to the same degree. Much depends on the types of foods you consume while using stevia. Diets high in stevia and low in carbohydrate are especially vulnerable to lean muscle loss. If you boost carbohydrate or sugar consumption to prevent lean muscle loss, then you may be adding more calories and more carbs into your diet, which could derail your weight loss efforts.

THYROID FUNCTION

Some evidence indicates that when whole stevia leaf or stevia extract (stevioside) is fed to livestock it causes them to eat more and gain more weight, and reduces their blood glucose and thyroid hormone (T3) levels. In one study, for instance, chickens who were given the standard feed supplemented with 2 percent dried ground stevia leaf or stevia extract (stevioside) were compared with chickens fed without supplementation. Both the stevia leaf and extract caused the animals to eat more and gain more weight, increased their abdominal fat, and reduced their blood glucose levels (an effect of elevated insulin) and their thyroid hormone (T3) levels.[27]

This study, published in 2008, is the first one I've seen that has reported any effect on thyroid function. Since lower thyroid hormone levels slow down metabolism, this could be another contributing factor to the reported increase in weight and body fat—something to be concerned about for those who are using stevia for weight loss. To date, I have not seen any other reports showing a relationship between stevia consumption and thyroid levels. Obviously, further research needs to be done to clarify this issue.

LONG HISTORY OF USE

The argument that "stevia has been in use for centuries," has often been cited as evidence of its safety. However, just because it has been used for a long time does not necessarily make it safe. While long use is evidence of a sort, there are examples of substances that were in wide use for centuries and even millennia—such as tobacco, cocaine, and even sugar—before their profound toxicity became apparent.

People have been using tobacco for centuries, and some even boasted that it was good for one's health. Its detrimental effects were hotly debated in the mid-20th century until it was conclusively proven to cause cancer and heart disease. The tobacco industry fought bitterly to hide its detrimental effects. They sponsored the publication of favorable studies, hid evidence, and paid doctors to be pitchmen for their products. It took the accumulation of a mountain of evidence to prove its harmful effects beyond a shadow of a doubt. In the meantime, many people's health failed because they continued to smoke or were exposed to second-hand smoke.

The leaves of the coca plant, another South American herb like stevia, have been used for thousands of years. The ancient Incas used to chew coca leaves to improve stamina and endurance—perceived health benefits of the plant. In 1859 chemists learned how to extract cocaine from the leaves and purify it into a white crystalline powder. From the 1860s to the early 1900s cocaine was used in elixirs, tonics, and wines. It was used by people of all social classes, including such notable figures as Sigmund Freud and Thomas Edison. It was considered so harmless that it was added to soda pop, as one of the original ingredients in Coca-Cola. In

Cocaine is an extract from the coca leaf, an herb that has been used for thousands of years.

1912 a reported 5,000 cocaine related deaths were recorded in the United States for that year. Yet, it wasn't until 1922 that the drug was officially banned.

Refined sugar has been around for about 1,000 years. During all that time no one considered it harmful. Today there is widespread concern about the potential dangers of excessive sugar consumption, which is the main reason stevia and other sugar substitutes have become so popular.

IS THERE A STEVIA CONSPIRACY?

People who use stevia assume that it is completely safe, a harmless extract from an herb. How bad can that be? They mistakenly believe there are no reliable studies that demonstrate any harmful effects. So why, they wonder, did the FDA put a ban on a seemingly harmless product? Some subscribe to a myth floating around on the Internet that the FDA, under pressure from food manufacturers (presumably the NutraSweet Company), put the ban in place to protect the profits of the companies that sold artificial sweeteners. It is all a grand conspiracy to keep a healthy natural sweetener away from the public! This story might sound good for a mystery novel, but it doesn't hold up in real life.

The conspiracy story surfaced soon after a series of incidents involving the FDA and the Stevita Company of Arlington, Texas. At the time, stevia was still banned as a sweetener and food additive, but was allowed as a dietary supplement. In 1996 the Stevita Company received their first shipment of their new stevia sweetener, Stevia Spoonful, packaged under the trade name Steviasweet. It was apparent that the product was intended to be sold as a sweetener rather than as a supplement, so the FDA detained the shipment stating that the trade name "Steviasweet" implied it was intended to be sold as a sweetener. Employees from the company went in and changed the name on each package to "Stevita," and the FDA released the product. This incident brought the company under the watchful eye of the FDA. The following year, 1997, the company began importing Stevita coffee flavoring—and again they got into trouble with the FDA, and again had to change the name on the packages. Later that year the company began selling stevia cookbooks that described how to use stevia as a sugar substitute.

The books made unauthorized health claims about stevia, and while that in itself did not break any laws, it is unlawful to use literature that makes unauthorized health claims about a particular product in order to sell that product. The FDA ordered the company to stop selling the books and to destroy them. From this arose the rumor that the FDA was "burning" stevia cookbooks, although no books were actually burned; soon thereafter a conspiracy theory was born. According to the theory, the FDA was in cahoots with the artificial sweetener companies to prevent competition from stevia. The most likely company behind it all was the NutraSweet Company, which supposedly bribed FDA officials to go after companies selling stevia. A couple of other companies also received warnings from the FDA. However, the vast majority of companies that sold stevia at that time, without breaking any regulations, were never harassed by the FDA.

Back in 1991, when the ban first went into place, the only artificial sweeteners available were aspartame, saccharin, and the relatively unheard of acesulfame potassium. Only 12.7 percent of the US population even used artificial sweeteners at the time.[28] Most of the sweeteners were used in little packets and in some soft drinks. Sugar substitutes were not the big businesses they are today.

Aspartame was developed in 1965 by a chemist working for G.D. Searle & Company. The sweetener was approved for use in the US in the early 1980s. In 1985, Monsanto Company bought G.D. Searle, and aspartame became a separate Monsanto subsidiary called the NutraSweet Company. The European patent on aspartame expired in 1987 and the US patent expired in 1992, allowing other companies to make the sweetener and compete with the NutraSweet Company. Since the patent expired in 1992, one year after the FDA's ban on stevia was put in place, it makes no sense that the NutraSweet Company would risk breaking the law to bribe government officials if they were going to lose their market advantage the following year. And the ban on stevia in Europe didn't go into effect until 1999, long after the aspartame patent expired.

Studies with laboratory rats during the early 1970s linked saccharin with bladder cancer. As a consequence, all foods containing saccharin were labeled with a health warning, that remained in effect until 2000. In the 1990s saccharin was sold almost entirely in little packets, and marketed specifically for people who could not use other sweeteners. It even carried a warning that it might promote cancer. Fear of cancer kept most people from using the product. So its market share was tiny compared to other sweeteners. Saccharin was developed in the 1879, so there is no patent protection on it. Again, it makes no sense that a company would try to suppress stevia to protect the sales of saccharin.

The only other artificial sweetener available in 1991 was acesulfame potassium. It was developed by Hoechst, a German company. The sweetener was approved in the US in 1988, but had a very small market share during the 1990s. Legalizing stevia would have had no impact on the sales of acesulfame potassium.

These manufacturers had no interest in suppressing stevia; indeed, they would have had more interest in exploiting it, as is done with any product perceived by the public as "natural" or "harmless." Such a sweetener could have been a goldmine for them. While whole natural products that have functional uses cannot be patented, derivative products can. Stevia extracts can be purified into patented products. In fact, it is much easier to use a natural product that has proven to have a certain characteristic and patent extracts or derivatives of it, than it is to create a synthetic product with the same effects. Stevia, rather than be viewed as competition, would be exploited as it has been since its approval in 2008. The conspiracy story is just a myth.

FDA APPROVAL

Stevia remained banned until the passage of the Dietary Supplement Health and Education Act of 1994, which specifically exempted products meeting the definition of "dietary supplement" from the FDA's food additive policies. As a result, stevia was allowed back in the US, provided it was clearly labeled as a supplement. This allowed marketers to sell stevia leaf and extracts as dietary supplements, but not as sweeteners or food additives. Regardless of how the product was labeled, people used the "supplement" as a sweetener anyway.

Since stevia was perceived as a healthier option to artificial sweeteners, its popularity grew. Major food and beverage producers such as Cargill and Coca-Cola took notice, and began developing stevia-based sweeteners. The sweetening agents in stevia were identified and purified. The most potent, commercially viable sweetening agent in stevia, rebaudioside A, was extracted, refined, and purified.

Cargill teamed up with Coca-Cola to develop a stevia-based sugar substitute marketed as Truvia. Not to be outdone, PepsiCo partnered with artificial sweetener maker Merisant, which produces Equal and Canderel sweeteners, to develop their own brand of stevia called PureVia. Merisant was originally formed from Monsanto's tabletop sweetener business, which was later sold to MacAndrews and Forbes, a multibillion-dollar holding company. To distance themselves from Monsanto and their own artificial sweetener brands, Merisant formed a subsidiary company with a wholesome sounding name, Whole Earth Sweetener Company, solely to produce and sell PureVia.

These companies spent millions of dollars developing these sweeteners and building marketing plans. However, if the FDA maintained the ban on stevia, their multimillion-dollar investments would be lost. In order to get the FDA to change its position on stevia, these companies had to sponsor studies and build a scientific database to use as ammunition to achieve their goals. These studies didn't have to prove that stevia itself was harmless or safe, they only had to show that rebaudioside A was. Starting in the early 2000s many studies were commissioned to prove the safety of rebaudioside A and other stevia extracts. One of the problems with research that is motivated by financial gain is the pressure on researchers to give backers what they want, regardless of the actual results. These studies, although favorable to rebaudioside A and other stevia extracts, must be viewed with some degree of skepticism.

With the financial backing of Cargill, Coca-Cola, PepsiCo, and other industrial giants, in just a few short years dozens of studies appeared in medical journals touting the safety of stevia and its possible health benefits. These studies refuted and attempted to discredit older, negative studies. Although a few studies showing

Truvia was developed by Cargill and Coca-Cola. PureVia was developed by PepsiCo and Merisant.

negative effects were also being published, their numbers were drowned out by this new chorus of positive research.

In 2008, with these favorable studies in hand, the makers of Truvia and PureVia approached the FDA to request GRAS status, which would enable them to sell their products as sweeteners and food additives. After the FDA reviewed their petitions, rebaudioside A was granted GRAS status, and in December 2008 it was pronounced safe to use as a food additive. This came as a surprise to many, since stevia had been restricted for so long. The FDA clarified the reason for their decision by stating that rebaudioside A is *not* stevia; because it is a highly purified product, it is entirely different.

Nearly all of the stevia studies published since the mid 2000s have been conducted to exonerate steviol glycosides, but not stevia leaf, from any wrongdoing. All of the concerns about stevia leaf are still a concern, and the FDA's 2008 ruling did not change its position on stevia itself. "Rebaudioside A is different than whole-leaf stevia or [other] stevia extracts, which can only be sold as dietary supplements," FDA spokesperson Michael Herndon has said. "Nobody has provided the FDA with evidence that whole-leaf stevia is safe." So the FDA's ban on other forms of stevia is still in place. They can be sold as dietary supplements, but cannot be labeled as sweeteners or sold as food additives.

Over the next few years the ban on steviol glycosides was lifted in Europe, Canada, Australia, New Zealand, and other countries, although the ban on the leaf is still in force in many of these countries.

IS STEVIA SAFE?

How safe or unsafe is stevia? Unless someone is allergic to it, stevia appears not to cause any acute or immediate problems, but the same thing can be said of sugar or soy or aspartame or sucralose. Studies of chronic use over a period of years have not been done. So we don't know how safe it really is. Problems with aspartame and sucralose didn't surface until years after FDA approval.

If stevia was as benign as its promoters claim, there would not be so many studies to the contrary. Promoters often claim that there have been no known adverse effects reported. But how can you tell if you are not producing a normal amount of sperm? How can you tell if your liver or kidneys are slowly degenerating? How do you know if you are gradually losing lean muscle mass, or producing less thyroid hormone, or if genetic mutations are occurring on a cellular level? These types of conditions don't usually manifest themselves as clear symptoms of ill health and, if noticed, are generally attributed to normal aging—certainly not to an herbal sweetener perceived as "healthy." The FDA's concerns about stevia's safety are legitimate.

One thing is very clear: stevia's wholesome image as a harmless and even health-promoting substance is not justified. While it may have little acute toxicity in small doses, in large doses it is highly toxic and can even be deadly. Studies show that *single* doses of about 80 times that suggested as acceptable for human daily consumption can cause death.[29] While few people will consume 80 times what you would get in a couple of sugar-free sodas, some people drink a lot more than two sodas a day, and often consume other foods sweetened with stevia on a daily basis. When a substance is approved as a food additive, it can find its way into thousands of food products. As the popularity of stevia grows, it will creep into more and more prepared foods. Total consumption can become significant, especially when consumed daily over a long period. This may lead to serious adverse effects beyond just obesity and insulin resistance.

5

Conflicting and Confusing Studies

The safety issue regarding stevia isn't cut and dried. As you've seen in the previous chapter, a number of studies by different investigators have raised legitimate concerns about the safety and usefulness of both stevia leaf and its purified extracts. Since about 2006 a plethora of new studies on stevia have appeared in the medical literature, almost all of which view stevia in a positive light. At this point, there are numerous studies that show stevia to be harmless or even beneficial. In fact, there are far more such studies available than there are those that suggest the opposite. If you were to argue for the use of stevia, you could choose from many studies to back your point of view. But that wouldn't negate the cautionary studies that have identified and even demonstrated potential problems. It also wouldn't prove that stevia is safe, or even that it is useful for its primary intended purpose as a weight loss aid.

Many, and maybe even most, of the conflicts in research results stem from legitimate differences in the design and execution of the studies. Results can be influenced by personal prejudice of the researchers, and by the subjects' age, gender, preexisting health issues, and dietary habits. Also, was the study done on rats, hamsters, guinea pigs, or humans? How long did the study last—a few days, weeks, months, or years? How much was used? Were they done *in vivo* (on living subjects) or *in vitro* (on

tissue cultures)? Did the study use stevia leaf, crude water extract, purified stevioside or rebaudioside A, mixed steviol glycosides, or steviol? Were other products combined with the stevia, such as maltodextrin, dextrose, or erythritol? How was the test substance administered? Was it delivered orally in water or a capsule, as a bolus inserted in the rectum, through a feeding tube going directly to the stomach, or injected into the bloodstream? These are all ways stevia has been tested.

Most people aren't aware of all these variables, and when they see a study on "stevia" they just assume it represents the sweeteners sold in the stores. Not necessarily. A stevioside extract injected directly into the bloodstream of a guinea pig may not produce the same results as a human consuming a sweetener made of rebaudioside A and maltodextrin. The results of any particular study may have no relationship to the effect stevia sweeteners have on us.

Some of the largest and richest food conglomerates in the world such as Cargill, Coca-Cola, PepsiCo, and Merisant, as well as pharmaceutical giant Johnson & Johnson are promoting and selling stevia-based sweeteners. Since FDA approval, sales of stevia-sweetened products are skyrocketing. Currently over 1,200 products are sweetened with stevia, making it a multimillion-dollar business.

One question you might ask is: why the dramatic increase in studies—the vast majority of them positive—in recent years? Would it have anything to do with Cargill and friends' entry into the stevia market, and their goals of getting FDA approval and influencing potential customers?

Big businesses use studies as marketing tools to sell products and make money. They are not interested in science or furthering some noble humanitarian cause. Their sole purpose in funding studies is to sell products. Positive studies are essential in getting FDA approval. They are also valuable in advertising and publicity promoting the benefits of a product. There is nothing necessarily wrong with this. The only problem is the accuracy of such studies. When businesses sponsor studies, researchers are under pressure to produce the desired results.

Researchers can design studies to get the results they want. For example, recent studies have shown stevioside to have no mutagenic activity. These studies have been used to argue against many of the earlier studies that have found mutagenic effects. These arguments, however, can be deceiving because in the digestive tract stevioside is metabolized to steviol, which has shown to induce gene mutations.[1] The claim that no studies have found stevioside to be mutagenic is technically true, but deceptive.

Let's look at another example. Concerns have been raised that stevia may cause hypoglycemia in some people. To examine this issue, studies have given rebaudioside A and/or stevioside to subjects in capsule form. Subjects show little or no effect on blood glucose, supposedly demonstrating stevia to be safe. But it is not stevia itself that affects blood sugar levels, it is the sweet taste of stevia that triggers the response. By giving subjects the stevia in capsule form, the sweet taste is bypassed, preventing the glucose-lowering effect.

These types of studies look convincing and can fool people who don't know the secret ways researchers sometimes manipulate their studies. Not all studies are purposely designed to get a predetermined outcome. Sometimes studies are influenced subconsciously by the researchers' own prejudices or preconceived beliefs. Expecting a certain outcome, they may subconsciously ignore or delete contrary data as insignificant or as random errors. This is so common that many studies are conducted under "double blind" conditions, which means that neither the researchers nor the subjects know if the subjects are taking the active test substance or an inactive placebo.

In 2010 researchers at University of California, San Francisco published a study that examined thousands of internal industry documents obtained as part of a number of legal settlements in the 1990s.[2] These documents which were turned over to the courts by the corporations involved, revealed the strategies they used to manipulate research and influence the scientific community and regulatory agencies. The paper outlined six main manipulative techniques:

1) Manipulate research to obtain predetermined results
2) Fund and publish research that supports industry interests
3) Suppress unfavorable research
4) Discredit researchers and research that do not support the industry's interests
5) Change or set scientific standards to serve corporate interests
6) Disseminate favorable research directly to policymakers (legislators and regulatory agencies), the media, physicians, and the public, bypassing the normal channels of scientific debate

These techniques were common among all the corporations involved in the lawsuits, perhaps because the marketing firms used by these companies employ the same time-tested strategies.

Does that mean we cannot trust any of the research being published in medical journals? No, but it does mean that you should be cautious about which journal articles to believe and which ones to question. How do you tell if a study is reliable or not? It is often difficult. One clue is to find out who benefits from the study. In other words, "follow the money." Which studies can be used in advertising and marketing to sell a product? Who stands to benefit most from the study's results? Who benefits from studies that claim stevia is safe or has some beneficial property? Let me list some: Cargill, Coca-Cola, PepsiCo, Merisant, and Johnson & Johnson—all huge international corporations which are involved in selling low-calorie sweeteners and various foods and beverages sweetened by them.

On the flip side, who stands to benefit from negative studies on stevia? Ahhh...nobody. You might think the sugar industry would

Businesses use studies as marketing tools to sell products.

benefit, but who controls the sugar industry? That's right: Cargill and friends. The same companies that sell stevia and artificial sweeteners also sell sugar. So revealing the possible dangers of stevia benefits no one. Therefore, the results of these studies are more likely to be reliable. Generally, these studies are being done by researchers who actually want to discover facts rather than promote the agenda of big business.

Just because you see a study that claims stevia is safe or has some medical benefit does not make it so. Nor does it erase all the studies that have raised concerns.

Conflicting studies are not isolated to just stevia; wherever money is on the line you will find disagreement in the medical literature. Industries sponsor studies to get favorable results to use in marketing and for government approval, and to sway medical and public opinion. Billions of dollars are on the line. Big corporations spend millions of dollars on studies to shape public opinion and deceive the medical community.

Any study that casts doubt on the safety or efficiency of a product that is bringing in millions of dollars in profit to its maker should be taken seriously and closely examined. The manufacturers of the product in question are generally quick to respond by sponsoring their own studies to discredit or counter the findings of the previous study. This tactic misleads the public and medical community, creates confusion and controversy, and takes the heat off the suspect product.

Artificial sweetener makers went ballistic when studies first started appearing claiming that aspartame and saccharin stimulated hunger leading to increased food intake. This was a threat to their profits! They weren't going to take this revelation lying down, and quickly fought back with their own company-sponsored studies to show the opposite. They sponsored numerous studies, far more than those demonstrating a relationship between artificial sweeteners and increased hunger—to give the impression that the overwhelming body of evidence was in their favor.

They did the same thing when it was discovered that artificial sweeteners promoted weight gain rather than weight loss. And the same thing again when some studies suggested a link between artificial sweeter use and cancer or neurologic problems.

Some company-sponsored studies have even suggested that artificial sweeteners are actually healthy. Many drugs have gone through this same cycle. The cholesterol-lowering statin drugs have been linked to a multitude of adverse health effects. For example, Lipitor, the best-selling statin drug, has been shown to cause memory loss and even amnesia. So the drug maker has produced studies to show Lipitor is actually *good* for brain health and can help *protect* against memory loss. These contradictory studies lead to a lot of confusion allowing manufacturers to continue to sell their products. Similarly, the tobacco industry spent millions producing studies that "proved" tobacco use does not cause heart disease and cancer; it took years of non-industry-sponsored research to counter the studies and propaganda produced by the tobacco industry. The more doubt manufacturers can create, the longer they can sell dangerous or ineffective products.

I suspect many people profiting off the sale of stevia products will object to the information I present in this book. They may even try to discredit it by citing studies that support their point of view. But as you have just seen, many of these studies are frauds produced for corporate profit.

6

Digestive Health and Function

GUT MICROBIOTA

The human gastrointestinal (GI) tract contains approximately 100 trillion microorganisms—bacteria, viruses, and yeasts—collectively known as the microbiota. It has been estimated that at least 10,000 and as many as 35,000 distinct species inhabit the human digestive tract. These organisms play an essential role in many aspects of human health. The human microbiota contain both beneficial or "friendly" microorganisms as well as not-so-friendly ones. The friendly bacteria perform many important functions essential for good health: they help maintain pH balance of the digestive tract, synthesize important vitamins such as vitamins B-12 and K, support immune function, aid in the breakdown and digestion of food, neutralize toxins, regulate glucose absorption, and much more.

Fortunately, they are more numerous than the potential troublemakers, and prevent them from overpopulating the digestive tract and causing harm. Disruption of this carefully balanced population of microbes has been implicated as a contributing factor in many health problems including obesity, insulin resistance and diabetes, reduced immune function, digestive disorders (chronic constipation, inflammatory bowel disease, Crohn's disease, celiac disease), neurological disorders (Alzheimer's, Parkinson's, autism, ADHD, depression), food allergies and sensitivities, eczema,

Human Gastrointestinal Tract

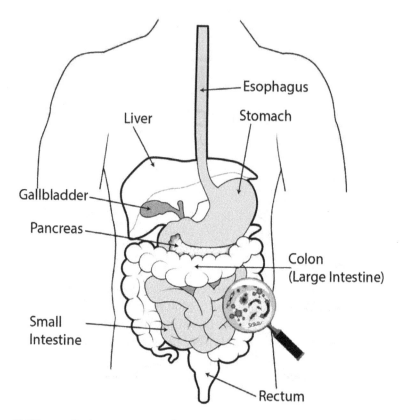

Trillions of microorganisms live in the GI tract.

recurrent yeast problems, and some forms of cancer. The health of our digestive system is so important that it has been said that up to 90 percent of all known human illnesses can be traced back to an unhealthy gut.

Many people are surprised to learn that the microbes inhabiting the gut can profoundly affect mental health as well. It seems that there is little connection between the gut and the brain, but scientists are just now learning about the intimate relationship between them. Next to the brain, the intestines contain the greatest number of neurons; some scientists have even referred to this mass of neurons as "the second brain." The vagus nerve, the longest of the 12 cranial nerves, is the primary conduit of information between hundreds of millions of nerve cells in your intestinal nervous system and your brain. The vagus nerve extends from the brain stem to the abdomen, controlling and monitoring many bodily processes associated with digestive function. Your microbiota directly affects the stimulation and function of the cells along the vagus nerve. Some of these microbes release chemical signals, just as neurons in the central nervous system do, to relay messages to the brain. Thus, the microorganisms in the gut can have a powerful influence on the brain.

In addition to housing nearly 3 pounds of bacteria and other microorganisms, the intestinal tract provides the mechanism by which the body extracts and absorbs nutrients from foods. The tissue lining the intestinal tract serves some very important functions. It provides a protective barrier between what is inside the intestines (food, bacteria, waste, chemicals, etc.) and the bloodstream or the rest of your body. It allows nutrients, such as vitamins, minerals, amino acids, and fatty acids, to pass into the bloodstream, but blocks the entrance of large food particles, bacteria, toxins, and other substances that may pose a threat to your health. The intestines contain a rich network of nerve tissues that constantly relay messages to and from the brain that control muscle movement that pushes food and waste along the digestive tract and monitors hormones involved in metabolism, satiety, and weight management.

Some of your friendly gut bacteria convert the fiber in your diet into short chain fatty acids (butyric and acetic acids), which are the primary food or fuel used by the cells lining your intestinal tract. These butyric- and acetic-producing bacteria are important to the health of your intestines and to the integrity of the intestinal wall; it is therefore important that our guts contain a healthy population of these bacteria. This is one of the reasons why nutritionists recommend we eat an adequate amount of foods rich in fiber.

The barrier between what is inside the digestive tract and the bloodstream generally consists of just a single layer of cells. The junction or tightness between the cells determines what can pass in or out. Nutrients pass through these junctions into the blood, but the space is too small to allow viruses or bacteria to enter. Conversely, water and mucus can flow *into* the intestinal tract, to aid in digestion and elimination.

The types of microorganisms living in the intestines exert a powerful influence on the permeability of these junctions. Too little of the good bacteria or too much of the wrong bacteria can damage the intestinal lining and increase the distance between its cells, making them more permeable. This is often referred to as a "leaky gut." When this happens, incompletely digested food particles can pass through the cell junctions into the bloodstream. If, for example, small fragments of protein enter the bloodstream, immune cells identify them as foreign particles, invaders, and trigger an immune response. This is how we get many of our food allergies. Bacteria, too, may pass through the compromised junctions causing chronic systemic inflammation, which greatly increases the risk of many health problems, including heart disease and diabetes.

Recent cutting-edge research has shown that the microbiota that populate the gut exert a tremendous influence on our health and are key to whether or not you live healthfully to a ripe old age. The most significant factor influencing the microbial populations within your gut is what you put into it—foods and drugs.

Antibiotics, for example, are designed to kill bacteria. While antibiotics may be necessary to overcome a serious infection, they generally kill much of the bacteria in the digestive system as well, allowing yeast (fungi) and viruses, which are not affected by antibiotics, to proliferate, disrupting your microbiome (the community of microorganisms in the gut). Steroids and other drugs can also negatively influence your microbiome. In contrast, probiotic supplements and fermented foods contain lactic acid–producing bacteria, such as Lactobacillus, that produce a slightly acidic environment that reduces the growth of less desirable microbes.

To maintain good health, then, it's important to have a healthy, properly balanced microbiota population. This is accomplished by eating healthy foods while avoiding certain foods, food additives, and medications that can disrupt proper microbiota balance. Consuming foods and drinks heavy in sugars and refined grains can change the microbiota population in an unhealthy way and promote weight gain. Many of the troublemaking gut microbes thrive on the sugar in your diet, enabling them to flourish. For this reason, many people have perceived sugar substitutes as a healthy alternative.

Most sugar substitutes are not well digested, and therefore provide little or no calories. This is why they are believed to be beneficial for weight loss. They go in and out of the body without contributing much, if anything, to the daily calorie load on the body. Gut bacteria cannot easily break them down or use them for food either, so they don't support the growth of unhealthy microbes the way sugar can. You get the sweet taste without the calorie kick or the spike in blood sugar. Unfortunately, they do something else that is not so sweet: they disrupt the normal gut populations that can throw your blood sugar out of balance and increase the risk of diabetes and obesity, as well as promote many other health problems.

NON-CALORIE SWEETENERS ALTER GUT MICROBIOME

One of the first clues that nonnutritive sweeteners might offset the delicate balance of the gut microbiome surfaced soon after the approval of sucralose (Splenda) in 1998. People who used it began to report adverse effects ranging from high blood pressure and dizziness to skin rashes and elevated blood sugar. But the most common complaint was digestive problems—sick stomach, bloating, and diarrhea.

Researchers at Duke University in North Carolina set out to find why so many people experience digestive disturbances when they consume Splenda. Their findings, published in 2008 in the *Journal of Toxicology and Environmental Health, Part A*, raised questions about the safety of the sweetener and set off a wave of protests from The Calorie Control Council (CCC), which promotes the use of artificial sweeteners, and McNeil Nutritionals, the maker of Splenda.

The Duke researchers evaluated the effects of Splenda on five groups of adult rats. In addition to their normal diet, one group was administered plain water, thereby acting as the control. The other four groups received different doses of Splenda in their water. Splenda is composed of the high-potency artificial sweetener sucralose (1.1 percent) and the fillers maltodextrin and glucose.

The doses used were 100, 300, 500, and 1,000 mg of Splenda per kg of body weight per day, which is equivalent to sucralose doses of 1.1, 3.3, 5.5, and 11 mg/kg/day. These dosage levels were selected because they span the range of values below and above the accepted daily intake for sucralose of 5 mg/kg/day established by the FDA.

The results were shocking. Even at doses within the FDA's acceptable daily intake, Splenda significantly altered the microbiome, disrupted pH balance, increased the expression of enzymes known to interfere with nutrient absorption, and caused weight gain. Here we have yet another study revealing that a non-caloric sweetener can promote weight gain. Plus, it appears that it also disrupts the microbiome and the normal function of the digestive tract.

After 12 weeks, half of the animals in each group were sacrificed and examined. Specific enzymes known to limit nutrient absorption—cytochrome P-450 (CYP) and P-glycoprotein (P-gp)—were measured. Both CYP and P-gp had increased by factors of 2.5 to 3.5. The number of beneficial bacteria had decreased by an incredible 50 percent relative to the control animals, with Bifidobacteria, Lactobacilli, and Bacteroides reduced by 37, 39, and 67.5 percent respectively. But there was no decrease in Enterobacteriaceae, a large family of bacteria that includes many familiar pathogens such as Salmonella, E coli, Klebsiella, and Shigella. The decrease in acid-producing bacteria, no doubt, affected the pH balance, making the colon more alkaline, promoting the growth of less desirable organisms. The body weight of the animals in all the groups increased during the study; even the animals that received less than the FDA acceptable limit of sucralose experienced more than a 100 percent increase in weight.[1]

All these dramatic changes occurred in just 12 weeks. If the same rate of change occurs in humans consuming acceptable doses of Splenda, their digestive environment (the microbiome) and function could be dramatically altered in a very short time.

Equally troubling is the fact that these adverse effects can persist for a long time, even after Splenda is discontinued. After the initial 12-week period, the animals were taken off of Splenda and given plain water with their diet for another 12 weeks. At the end of the second 12-week period these adverse effects remained. This suggests that if you use Splenda for a period of time and then quit, your gut microbiome will remain out of balance indefinitely, unless you do something to purposely rebalance it, such as eat probiotic supplements and fermented foods and reduce sugar intake.

Splenda proponents criticized the Duke study, saying the results were applicable only to rats, not to humans. Rats, however, are standard subjects for these types of studies because they respond very similarly to humans. Besides, of the 110 studies used to prove Splenda's safety in order to get FDA approval, the vast majority were done using animals. Only two studies used humans, involving a total of only 36 people. The longest of these human studies lasted just four days, and focused on Splenda's impact on tooth decay. The Duke University researchers also pointed out that the acceptable daily limit on Splenda approved by the FDA for humans was based on rat studies.

GLUCOSE METABOLISM

While it appeared that Splenda has a pronounced negative effect on the gut microbiome and digestive function, researchers next wanted to know if other non-caloric sweeteners had the same effect, and how altering the microbiome might affect glucose intolerance (insulin resistance). Glucose intolerance can lead to a host of health problems, including diabetes and Alzheimer's, and to an increased risk of liver, kidney, and heart disease. Researchers from the Weizmann Institute of Science in Israel set out to find the answer.

In the initial set of experiments, the Israeli scientists separated healthy 10-week-old mice into six separate groups. They added saccharin (the sweetener in the pink packets of Sweet'N Low), sucralose (the yellow packets of Splenda), or aspartame (the blue packets of Equal) to the drinking water of three groups of mice. The other three groups drank plain water or water supplemented

with either glucose or sucrose (table sugar). After 11 weeks, all of the mice that had consumed the non-caloric sweeteners developed marked insulin resistance, which is indicated by higher than normal blood glucose levels. However, none of the mice in the other three groups consuming water and sugar water displayed any sign of insulin resistance. Each of the non-caloric sweeteners produced a similar effect to each other, demonstrating a much greater influence in promoting glucose intolerance than sugar.

When the Israeli researchers treated the mice with antibiotics, killing the bacteria in the digestive system, blood sugar levels returned to normal. This experiment provided a strong indication that the gut bacteria were intimately involved in the disturbance of blood sugar regulation.

To further test their hypothesis that the change in glucose metabolism was caused by a change in bacteria, the Israeli researchers performed another series of experiments. DNA sequencing of fecal samples from saccharin-fed mice showed that saccharin markedly changed the variety of bacteria in the guts of these mice. Then the investigators took intestinal bacteria from mice that had been consuming saccharin-laced water and injected them into the digestive tracts of mice that had never been exposed to saccharin. The saccharin-free mice developed insulin resistance, again demonstrating the connection between gut bacteria and blood sugar control.

Next, the Israeli researchers focused their attention on the gut bacteria populations in human subjects. In the 381 nondiabetic participants in the study, the researchers found a correlation between the reported use of *any* kind of artificial sweetener and signs of insulin resistance and being overweight and different gut microbe profiles. Those who regularly consumed artificial sweeteners, and particularly those who consumed the highest amounts, showed higher fasting blood glucose levels, poorer glucose tolerance, and different gut microbe profiles compared to those who did not consume these sweeteners.

Finally, the researchers recruited seven lean, healthy human volunteers who normally did not use artificial sweeteners, and over six days gave them the maximum amount of saccharin recommended as safe by the FDA. Incredibly, in just six days, four of the seven healthy subjects' blood sugar levels indicated signs of insulin resistance, with abrupt changes in their gut microbes, in the same way as in the affected mice. The three volunteers whose glucose tolerance was not altered, showed no change in their gut microbes.[2] This last experiment only lasted six days; if it had continued, it's likely that all seven volunteers would have eventually developed altered gut microbiome and glucose intolerance.

Further, when the Israeli researchers injected the affected human participants' bacteria into the intestines of mice, the animals developed insulin resistance, demonstrating that the effect was the same in both mice and humans.

This study clearly shows that non-caloric sweeteners can disrupt the intestinal microbiota and the body's ability to regulate blood sugar, causing metabolic changes that can be a precursor to diabetes and obesity. These are "the very same conditions that we often aim to prevent" by consuming sweeteners instead of sugar, said Dr. Eran Elinav, an immunologist at the Weizmann Institute and one of co-authors of the study.

"Here we demonstrate that consumption of commonly used non-caloric artificial sweetener formulations drives the development of glucose intolerance through induction of compositional and functional alterations to the intestinal microbiota," said the researchers.

It appears that non-caloric sweeteners can make us sick and fat. These studies are very compelling; enough so that we should think very seriously about using any non-caloric sweetener. Dr. Elinav said he has already changed his own behavior. "I've consumed very large amounts of coffee, and extensively used sweeteners, thinking like many other people that they are at least not harmful to me and perhaps even beneficial," he said. "Given the surprising results that we got in our study, I made a personal preference to stop using them." Smart move. It would be wise for everyone to stop using low-calorie sweeteners.

It's important to note that the Israeli researchers used three different non-caloric sweeteners in this study: saccharin, sucralose, and aspartame. Each sweetener is chemically very different. So it wasn't the unique character of chemicals themselves that caused the observed effects. The only similarity between the sweeteners is the sweet taste without calories. All of the sweeteners caused changes in the intestinal bacteria population that led to pronounced metabolic changes that altered the gut microbiome, leading in turn to insulin resistance. This study sounded the alarm that perhaps *all* non-caloric sweeteners may be contributing to our growing epidemic of diabetes and obesity, including stevia.

Yes, even stevia can have an unwanted effect on our microbiome. *Lactobacillus reuteri* is a major beneficial strain of bacteria that inhabit our GI tract; they produce lactic acid, and are the primary bacteria in fermented foods. They are also the major component of probiotic supplements. Interestingly, studies have shown that stevioside and rebaudioside A inhibit the growth of *Lactobacillus reuteri* bacteria, yet not the growth of some of the more troublesome bacteria species.[3] This suggests that stevia may be an anti-probiotic, meaning it decreases beneficial lactic acid gut bacteria and promotes an unhealthy microbiome.

DIABETES

Insulin resistance is the hallmark feature of type 2 diabetes. Severe insulin resistance leads to diabetes, a condition in which blood sugar levels are elevated all the time. When blood sugar is raised, insulin is pumped into the bloodstream, raising blood insulin levels. Excessive sugar consumption has long been suspected as a leading cause of insulin resistance and diabetes because frequent consumption of sugar keeps blood sugar and insulin levels elevated for extended periods of time. This high exposure to insulin desensitizes the cells to the action of insulin, causing them to become insulin resistant.

Although zero-calorie sweeteners do not raise blood glucose levels like sugar does, they promote insulin resistance and diabetes even *more* than sugar does. A study by French researchers investigated the association between sugar-sweetened beverages, low-calorie sweetened beverages, or 100 percent fruit juice (with no added sweetener). A total of over 66,000 woman participated in the study, which extended over a 14-year period. The researchers found that both sugar sweetened and artificially sweetened beverages increased the risk of insulin resistance and type 2 diabetes. No such association was observed for the fruit juice consumption in this study, even though fruit juice contains a good deal of sugar.[4] While non-caloric sweeteners do not raise blood sugar levels immediately after consumption, they have a long-term effect that causes chronic elevated blood sugar. Elevated blood glucose causes excess insulin to be produced. Insulin is needed to remove the glucose from the bloodstream and shuttle it into the cells. But insulin also triggers the conversion of glucose into fat and shuttles it into fat cells. Since there is a correlation between diabetes and being overweight, this study provides yet further evidence that non-nutritive sweeteners promote weight gain.

Soda consumption has been linked with increased risk of diabetes, presumably because of the sugar content, however, studies indicate that artificially sweetened beverages are even worse, promoting diabetes more than sugar sweetened ones. This is definitely true for aspartame and other artificial sweeteners, but what about stevia? A study by researchers from Kanazawa Medical University in Japan investigated the association between the consumption of diet soda and the incidence of type 2 diabetes in Japanese men. Subjects were monitored for a period of seven years. In Japan most diet sodas are sweetened with stevia, so this study provided an evaluation of the effects of the long-term use of stevia. The researchers found a significant association between stevia sweetened diet sodas and an increased risk of diabetes in Japanese men and concluded that consuming diet soda in place of sugar sweetened soda is not effective at preventing type 2 diabetes.[5] I hate to busts people's love affair with stevia, but the available evidence suggests that it increases the risk of developing type 2 diabetes more than sugar.

Another study published in the journal *Nature* showed that virtually everyone who has type 2 diabetes has an abnormal microbiome—specifically, reduced numbers of butyrate-producing bacteria.[6] This is significant because butyrate (butyric acid) is the preferred source of energy for the cells lining the human digestive tract, and is necessary for the repair and maintenance of the intestinal wall. The study also found an increase in the number of potentially pathogenic organisms. Increased oxidative stress was also noted, implying that there is increased damage occurring to intestinal tissue.

MICROBIOME AND OBESITY

Do you have a little extra baggage on your hips, thighs, stomach, or butt? Have you struggled to lose weight on low-calorie diets without success? Do you eat low-fat, sugar-free foods sweetened with low-calorie sweeteners, and as hard as you try, the weight just doesn't want to come off? If so, you are one of millions of people who unknowingly sabotage your weight-loss efforts by using low-calorie sweeteners.

Investigators have discovered that the microbiome in overweight individuals is very different from that of those who are normal weight. Non-caloric sweeteners promote the growth of bacteria that are particularly abundant in overweight individuals.[7-8]

Jeffrey Gordon, a physician and biologist at Washington University in St. Louis, has done research showing that the correlation between bacteria and obesity is more than a coincidence. Gordon notes that more than 90 percent of the bacterial species in the gut come from just two subgroups—Bacteroidetes and Firmicutes. Several years ago Gordon and his team found that genetically obese mice (animals that lack the ability to make leptin, a hormone that limits appetite) had 50 percent fewer Bacteroidetes and 50 percent more Firmicutes bacteria than normal mice did. When they transferred a sample of the Firmicutes (fat-promoting) population from the obese mice into normal-weight ones, the normal mice became fatter. The reason for this response, Gordon says, was twofold: Firmicutes bacteria transplanted from the fat mice produced more of the enzymes that helped the animals extract and get more energy (calories) from their food, and the bacteria also manipulated the genes of the normal mice in ways that triggered the storage of fat rather than its breakdown for energy.[9] Bacteroidetes, on the other hand, are more adept at breaking down plant material and fiber into short chain fatty acids (butyric and acetic acids), which are so beneficial for the health and integrity of the intestinal lining.

A follow-up study by Dr. Gordon and his team was performed using human subjects—sets of twins, where one was obese and the other normal weight. The researchers transferred gut bacteria from obese human twins and put them into the gastrointestinal tracts of slender mice—and the mice became fat. When bacteria from the slender twins were introduced into lean mice, the mice stayed lean. These results demonstrated the powerful influence that the different populations of gut microbiota can have on body weight.[10]

Obese people have a greater amount of Firmicutes bacteria than normal weight people. Gordon found that when people lose weight using either a low-fat or low-carbohydrate diet, the proportion of Bacteroidetes to Firmicutes bacteria increases. Stanford University microbiologist David Relman says this finding suggests that the bacteria in the human gut may not only influence

our ability to extract calories and store energy from our diet, but also have an impact on the balance of hormones, such as leptin, that shapes our eating behavior, leading some of us to eat more than others.

A number of studies have now shown that the composition of the intestinal microbiome can contribute to the progression of obesity.[11-13] Epidemiological studies in humans have shown that antibiotic treatment during the first six months of life, or disrupted colonization of the digestive tract due to caesarian section delivery, can increase the risk of being overweight later in life.[14-15] These two situations have no direct connection to caloric intake or metabolism, but have significant effects on the microbiome, and consequently body weight. A high-carb or high-sugar and low-fiber diet can also influence the composition of the gut microbiome. Low metabolism can intensify the above effects (see chart on page 81).

The ratio of Firmicutes to Bacteroidetes, is now considered a biomarker for obesity. In other words, the more Firmicutes you have in comparison to Bacteroidetes, the higher your risk of obesity. But that isn't all: larger proportions of Firmicutes also alter hormone regulation and increase the risk of systemic inflammation, diabetes, heart disease, dementia, and food allergies.

While not all Firmicutes are necessarily harmful, this group of bacteria includes a number of undesirable bacteria such as Streptococcus, Staphylococcus, Listeria, and Clostridium, which can stimulate inflammation within the intestinal tract, weaken the intestinal wall (promoting a leaky gut), interfere with digestive function, and cause disease.

SUGAR SUBSTITUTES LINKED TO OVERWEIGHT CHILDREN

The incidence of childhood obesity has more than doubled in the last 30 years. One-third of children in developed countries are now overweight or obese,[16] putting them at increased risk for developing cardiovascular disease, diabetes, and mental health disorders. More than 20 percent of preschool age children are classified as overweight or obese.[17] The consumption of low-calorie sweeteners during childhood has contributed to this growing problem.

A study by Canadian researchers has now discovered that women who consume low-calorie sweeteners while pregnant increase their risk of having overweight or obese children.[18] Animal studies have shown that the consumption of artificial sweeteners during pregnancy may predispose offspring to develop obesity and metabolic syndrome.[19] Human studies have documented an increase in premature births, allergies, and forearm fractures due

Pathways In Microbe-Induced Obesity

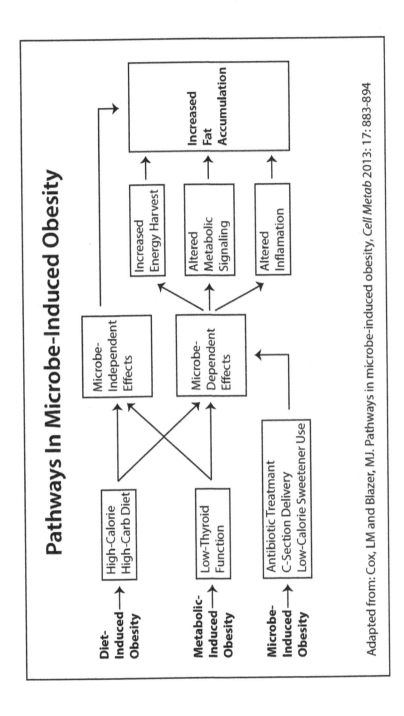

Adapted from: Cox, LM and Blazer, MJ. Pathways in microbe-induced obesity. *Cell Metab* 2013: 17: 883-894

to artificial sweetener use during pregnancy.[20-22] This study was the first to evaluate weight in humans exposed to artificial sweeteners in utero.

The Canadian researchers studied 3,033 mothers who had delivered healthy, normal-weight babies, and examined the children at one year of age. Thirty percent of the women had consumed artificially sweetened beverages while pregnant. After controlling for maternal body mass index, age, breastfeeding duration, maternal smoking, maternal diabetes, timing of the introduction of solid foods, and other factors, they found that compared with women who drank no diet beverages, those who drank, on average, one can of diet soda per day doubled their risk of having an overweight one-year-old. Since there was no association with infant birth weight and diet soda consumption in this study, the effect was not attributed to fetal growth. In contrast, the consumption of sugar-sweetened drinks was not associated with an increased risk of overweight one-year-olds.

It is interesting that the infants examined in this study were all normal weight at birth. So their abnormal weight gain occurred afterwards. How could diet soda consumption have influenced postnatal weight gain but not fetal growth? The authors of the study suggest that the effect was the result of the mother passing on to her child the same type of gut microbiota as she had herself. Infants are born with a sterile digestive tract. As the infant passes through the birth canal it is exposed to and picks up the mother's bacteria. This is the bacteria that will begin to inhabit the infant's digestive tract. If the mother has a microbiome that is out of balance, with more Firmicutes than Bacteroidetes, the same proportions of each will develop in the infant, predisposing the child to obesity and metabolic problems.

A good diet and exposure to a variety of bacteria in the child's environment may correct this problem in time. But the initial imbalance in the microbiome may instead be reinforced if the mother continues to consume artificial sweeteners while nursing. These sweeteners pass to the infant through the mother's milk, continuing to upset the infants' microbiome. If this is indeed the case, then any mother who uses artificial sweeteners while nursing is increasing her child's risk for obesity and other health problems. All low-calorie sweeteners, including stevia, are linked to alterations in the gut microbiome.

METABOLIC SYNDROME

Metabolic syndrome refers to a cluster of conditions that occur together that greatly increase the risk of heart disease, stroke, and diabetes; and makes a person more susceptible to other health problems such as polycystic ovary syndrome, fatty liver, gallstones, asthma, sleep disturbances, and some forms of cancer. The conditions that define metabolic syndrome include high blood pressure, elevated blood sugar levels (insulin resistance), excess body fat around the waist, and abnormal cholesterol levels (low HDL, high triglycerides). Abdominal obesity is perhaps the most significant of all these conditions. Having just one of these conditions doesn't mean a person has metabolic syndrome. However, any of these conditions increase your likelihood of experiencing health problems, and the more of them you have, the greater the risk.

Non-caloric sweeteners alter the gut microbiota in such a way as to increase body fat, particularly abdominal fat, and to increase the risk of metabolic syndrome and type 2 diabetes. A study conducted by a team of researchers from the United States and Europe found that the daily consumption of diet soda significantly increases abdominal obesity, and is associated with a 36 percent greater risk of metabolic syndrome and a 67 percent greater risk of type 2 diabetes compared with not drinking diet soda.[23]

Another study evaluated nearly 10,000 subjects and found a positive correlation between the consumption of diet soda and the incidence of metabolic syndrome. They found no such correlation with sugar sweetened beverages.[24] Interestingly, the consumption of milk slightly decreased the risk of developing metabolic syndrome, perhaps because people who drink milk are less likely to drink diet soda.

SATIETY HORMONES

As a source of various regulatory hormones, the GI tract is believed to play an important role in regulating the appetite.

Postprandial (after eating) satiety is believed to be regulated by a sensory system that communicates between the gut and appetite-regulating centers in the brain. In the gut there exists a suite of endocrine cells that synthesize and release various hormones in response to nutrient and energy intake, and it has been demonstrated that these hormones influence appetite in both humans and animals.

In the GI tract, glucose triggers the release of satiety hormones that tell the brain it's time to stop eating, so that excess calories are less likely to be consumed. However, artificial sweeteners, including stevia, do not trigger the release of these satiety hormones.[25] Therefore, when you eat foods containing non-caloric sweeteners, satiety is delayed and overeating can become an issue.

If you are overweight, the problem can be compounded. There is evidence suggesting that satiety hormone responses are impaired in obese individuals.[26] It's possible that the change in gut microbiota to increasing numbers of Firmicutes and decreasing Bacteroidetes delays the release of these hormones and consequently the feeling of satiety.[27]

YOUR INTESTINES CAN TASTE SUGAR

Taste is the sensation produced when a substance activates taste receptor cells found in the surface regions in the mouth. When these receptors are activated, electrical impulses are transmitted to the brain. The human body recognizes five taste qualities: sweet, salty, bitter, sour, and umami. Taste receptors in the mouth function as gatekeepers for the digestive system to ensure that we consume essential nutrients for survival and health, while rejecting potentially harmful or toxic foods. For example, a salty taste signals the presence of either sodium or minerals; umami indicates the presence of proteins; excessive sour taste signals spoiled food; bitter taste often indicates the presence of poisons; and sweet taste indicates the presence of carbohydrates or energy producing nutrients. The ability to taste and identify foods that are nutrient-rich while avoiding toxic substances has been essential for our survival throughout the course of human history.

Taste receptors do far more than just stimulate sensations of delight as you enjoy a delicious meal, they kick off a series

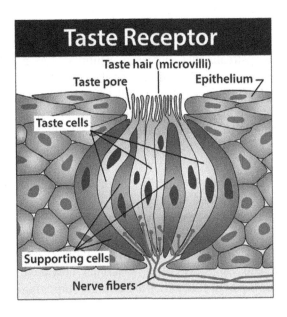

Taste Receptor

Taste hair (microvilli)

Taste pore

Epithelium

Taste cells

Supporting cells

Nerve fibers

of chemical reactions that affect your physiology as well as influence the types of bacteria living in your digestive tract. The sweet taste of non-caloric sweeteners, combined with the lack of corresponding calories, can alter the balance between Firmicutes and Bacteroidetes. Why do non-caloric sweeteners promote the proliferation of Firmicutes and the decline of Bacteroidetes? The answer to that question was provided, in part, by a discovery that identified taste receptors inside the gastrointestinal (GI) tract. Just like your tongue, your throat, stomach, and even your intestines have taste receptors that can sense sweet, savory, and bitter foods. Your intestines can literally taste the sweetness of sugar. While you don't consciously notice these tastes as food travels down your digestive tract, your brain and intestines do.

The discovery of taste receptors in the GI tract explains a phenomenon that has mystified physiologists for more than 50 years, known as the "incretin effect." Incretins are gut hormones that stimulate pancreatic cells to secrete insulin. The incretin effect was first described in the 1960s, and refers to the fact that eating glucose triggers significantly greater insulin response than an intravenous injection of glucose, even when the doses are matched to cause the same increase in blood glucose levels. Researchers observed that oral glucose was inducing the release of incretins into the bloodstream that increased insulin secretion more than the glucose injection. Neuroscientist Robert Margolskee, MD, PhD, now the director of the Monell Chemical Senses Center in Philadelphia, Pennsylvania, realized that if there were receptors in the intestine that could detect glucose and trigger the release of these hormones, this would provide the missing link for the incretin effect. In 2007, his hypothesis proved correct as his team located the sweet receptor cells within the intestinal wall.[28] Since then taste sensors have been found throughout the GI tract.

Although taste receptors seem out of place anywhere other than the mouth, this is only because they were first found in taste buds on the tongue. Taste receptors are simply a way of sensing chemicals, and can have other functions unrelated to simply detecting the flavor of food. They are actually surprisingly common in the body, and their presence in some locations is sometimes baffling to scientists.

Sweet-taste sensors are activated not only by sugar, but by anything sweet, including non-caloric sweeteners. "We now know that the receptors that sense sugar and artificial sweeteners are not limited to the tongue," says Margolskee. "Cells of the gut taste glucose through the same mechanisms used by taste cells of the tongue. The gut taste cells regulate the secretion of insulin and hormones that regulate appetite. Our work sheds new light on how we regulate sugar uptake from our diets and regulate blood sugar levels."

The sweet taste of sugar in the GI tract triggers the release of several hormones that regulate insulin secretion, glucose absorption, appetite and calorie intake, and metabolism.[29]

Non-caloric sweeteners, which are not absorbed or metabolized, strongly stimulate glucose receptors.[30] Unlike glucose, which is absorbed and removed from the digestive tract, rebaudioside A, sucralose, and other nonnutritive sweeteners are not easily broken down or absorbed, but remain in the GI tract. As they journey through the GI tract they continually stimulate sweet-taste sensors. Therefore, they have a far greater effect on sweet-taste signaling than glucose does.

Let me make an analogy that will illustrate what is going on here. It's like a molecule of glucose coming to the door of the intestinal wall and pressing the doorbell (taste receptor). The door opens and it enters (is absorbed into the bloodstream). That's the normal process. It's different with fake sweeteners. When rebaudioside A, erythritol, sucralose, or any other nonnutritive sweetener comes to the door and rings the bell, the door opens but the sweetener can't fit through the doorway (it is not absorbed but remains in the GI tract), so it keeps ringing the doorbell again and again and again for 24 hours or more. Every time the doorbell is rung, taste sensors are activated and hormones are released. This repeated activation of the taste sensors produces a flood of hormones and the digestive tract is primed to absorb glucose, a lot of glucose, but the expected glucose is not there. This leads to chemical changes in the intestines that promote the growth of Firmicutes (the obesity microbes) and stifles the growth of Bacteroidetes (anti-obesity microbes).

The overactivation of the sweet-taste sensory cells by repeated use of non-calorie sweeteners may very well cause them to burn out and become dysfunctional, which would seriously disrupt glucose metabolism, leading to an assortment of metabolic problems that could include obesity, diabetes, inflammatory bowel disease, and heart disease.[31-32]

GASTROINTESTINAL EXCITOTOXINS

In the GI tract, non-caloric sweeteners function as excitotoxins—substances that can cause the overstimulation of sensory cells, leading to their death. A very similar effect is seen in the brain as a consequence of excessive exposure to certain amino acids that function as neurotransmitters. The two most common of these are glutamate and aspartate. Our greatest exposure to these two excitotoxins comes from the flavor enhancer monosodium glutamate (MSG) and the artificial sweetener aspartame.

The term "excitotoxicity" was coined by neuropathologist John Olney in 1969 after he observed that feeding monosodium glutamate to newborn mice destroyed neurons throughout their brains. The term was used to describe the destructive activity caused by the overexposure of brain cells (neurons) to glutamate, aspartate, phenylalanine, and other excitatory neurotransmitters. While neurotransmitters are a normal and even essential component of brain communication, exposure to too many of them at one time can create a state of toxicity, and in this situation these neurotransmitters become what is referred to as *excitotoxins*.

The brain uses neurotransmitters to relay messages from one neuron to another. Glutamate and aspartate are excitatory neurotransmitters, meaning they stimulate chemical and electrical activity within the neuron that eats up the cell's energy. Exposure to too many excitatory neurotransmitters at one time can overstimulate the neurons, driving them into a frenzied state of overactivity, exhausting the neuron's energy reserves, causing them to burn out and die. During this process, a large number of destructive free radicals are generated that promote inflammation and cellular damage, compounding the problem. A growing number of studies are linking the excessive exposure to dietary glutamate and aspartate to neurodegenerative diseases such as Alzheimer's, Parkinson's, ALS, and others. Even typical memory loss, mild intellectual deterioration, and loss of coordination that frequently occurs in late middle age may be related to excessive consumption of excitotoxins.

Glutamate and aspartate are commonly found in foods. However, eating foods containing these amino acids is not a problem. What is a problem is when they are purified and concentrated into MSG, aspartame, and similar food additives, at which point they become troublesome. Glutamate and aspartate found naturally in foods are always bound to other amino acids. The process of breaking the bonds and releasing individual amino acids takes time, so the amino acids are released slowly. Thus, blood levels of glutamate and aspartate are kept within reasonable bounds, which the body is capable of handling. Frequently eating foods containing MSG and aspartame exposes the brain to continuously high levels of these excitotoxins.

Eating foods containing nonnutritive sweeteners have a similar effect in the digestive tract. Non-caloric sweeteners are typically 100 to 600 times sweeter than sugar, so they can deliver a tremendous wallop to the taste receptors in the gastrointestinal tract. Since most of these sweeteners are not easily broken down, they repeatedly activate taste receptors all along the digestive tract. While taste receptor cells are not neurons, they are connected to neurons that relay signals to the brain. Overstimulation can burn out both the taste receptors and their associated neurons. Stevia, or rather steviol glycosides, act as gastrointestinal excitotoxins that can burn out taste receptors, release free radicals, initiate inflammation, disturb glucose metabolism, change pH, and alter the gut microbiota population.

If you would like to learn more about excitotoxins I highly recommend the book by neurosurgeon Russell Blaylock, MD, *Excitotoxins: The Taste That Kills*.

7

Adverse Effects

REPORTED SIDE EFFECTS

Stevia promoters like to point out that the sweetener has been used in Japan, Taiwan, Korea, Paraguay, Brazil, and Israel for many years without any apparent harm. "Apparent harm" means that there have been no obvious connections between stevia use and any major health problem such as cancer, seizures, kidney failure, and the like. But that doesn't mean stevia is harmless or that it isn't associated with other, less obvious, effects.

Aspartame, saccharin, sucralose, and even sugar have also been used for many years without any apparent harm. That is why many medical professionals claim they are perfectly safe. However, there are studies that suggest that these products may indeed have harmful effects, and people have reported adverse effects from using them. For these reasons, many people believe these sweeteners are unhealthy.

Likewise, there are many studies that suggest stevia may cause health problems and many people do experience adverse side effects with its use. Promoters claim that stevia is totally harmless and that nobody suffers any harm from its use. After all, they say, it's just an herb, and how can an herb be harmful? Although you don't often hear about adverse effects from stevia, they have been reported by quite a number of users. Human clinical studies have also reported various side effects from consuming stevia. Some of the documented side effects mentioned in published medical studies include nausea, abdominal discomfort, muscle pain, headache, fatigue, and dizziness.

In one double blind study, 13 percent of 60 subjects taking an extract of purified steviol glycosides experienced side effects troubling enough to report to the investigators; in three cases the effects were so severe the subjects were forced to withdraw from the study.[1] According to the data in this study, one out of every five people who use stevia may suffer some type of noticeable adverse reaction.

Other, more subtle effects, such as a gradual change in the population of gut microbiota or increasing glucose intolerance, may not be immediately noticeable and may affect essentially all regular users. Many users may not recognize noticeable symptoms as being caused by stevia because of the general belief that stevia is harmless, it couldn't possibly be the cause of their pain or discomfort; instead, they may attribute their symptoms to something else, such as aging or stress.

Because of the general belief that stevia is a harmless, natural product, there is resistance to the very thought that it can possibly cause adverse side effects. I myself refused to believe stevia was the cause of side effects I was seeing in myself and others for a long time. Some people even get upset or angry if someone speaks negatively about their beloved stevia; it's almost as if one of their family members was being insulted. A few people, however, do make the connection; they notice that symptoms occur when they consume stevia and disappear when they don't.

An Internet search reveals that adverse side effects from stevia consumption are much more common than you might expect. The most common symptoms reported include abdominal pain, cramping, headaches, nausea, vomiting, dizziness, fatigue, muscle cramps, body aches and pain, throat irritation, phlegm, mouth sores, skin rash, and addiction. That's quite a number of reported side effects from an herb that is supposed to be completely harmless. These reactions include all of the side effects reported in clinical trials and then some. The following are some of the comments by stevia users.

Dermatitis

"I just learned that my body cannot break down stevia and it caused a HUGE crazy rash on my legs that itches beyond tolerance."

JJ

Joint Pain

"I have been using WalMart brand stevia every day, all day, for the last two years. I was convinced I might have rheumatoid arthritis. After wondering why I was experiencing muscle cramping and feet pain, I decided to look up stevia, as my last resort as to why I hurt. Sure enough, there are other stevia users out there with the exact pains I have been having."

Jessica

Digestive Distress

"I went from aspartame to Splenda in my soda, and then tried Coke Life with stevia when it came out. The stevia really screwed up my stomach. I quit all artificial sweeteners and will just treat myself with a real sugar Pepsi now and then."

Theresa

"I'm on a very strict diet because I am a fitness competitor. I know exactly every ingredient that goes into my body and when every day. These past two days I've switched from normal coffee creamer to using a bit of stevia in my coffee and have had cramps in my upper stomach all day long. I'm not a fan of artificial anything and have been eating clean for 10 weeks now so I can only attribute it to the stevia. Pretty scary. Definitely throwing it out tonight."

Selena

"When I tried stevia I got abdominal pain. At first I wasn't sure it was stevia but subsequently have definitely made the connection due to repeated episodes of the same. The more I take, the worse it is. Recently I used the liquid form and it was too much. The pain was hard to bear, like stabbing in the gut along with intense bloating. This all lasted over 24 hours. Truly awful. Never again. I would caution against it. I guess people are different, but for me it is clear . . . Many people react to this like I am accusing their mother of murder, 'Poor little stevia is harmless,' etc. Don't be fooled, stevia is capable of causing severe pain and who knows if it is causing damage as well."

George

"The worst of all is Truvia brand of stevia. The first time I ingested it, I had severe distress for over two weeks. Burning sensation in the mouth was the initial reaction, followed by muscle spasms and severe stomach ache and bloating. I had been eating a careful, limited diet of only unprocessed foods for a month when a friend served strawberries dipped in stevia. The first taste was bitter, followed by sweet, then the burning in mouth. Even my teeth hurt. I stopped eating, immediately spit out the food and rinsed my mouth . . . I then researched online, side effects from Truvia. There were multiple websites containing reports of side effects ranging from mild to severe. A number of people reported identical side effects to the ones I experienced. One woman, a tennis pro, had been unable to play tennis for months due to debilitating muscle spasms. Only after she discontinued Truvia as a sweetener did she recover. This stuff is dangerous!"

Kali

Mouth Sores

"I've been getting sores in my mouth, a raspy voice, and a few times have had difficulty swallowing. I've had a camera shoved down my throat and the doc thought acid reflux and told me to avoid coffee. When I quit putting stevia in my coffee for a week everything returned to normal."

Kym

"I was using Truvia for a while but it gave me awful stomach pain as well as canker sores."

Chris

Headaches and Dizziness

"When I used stevia I had headaches and body aches. I have decided to lay off all unnatural foods and drinks and I feel so much better."

Sandra

"Stevia can be toxic to some. It is a vasodilator, lowering blood pressure and lowers blood sugar. I have found that each time I have ingested stevia I got extremely dizzy and nauseated. If you don't need blood pressure lowered or blood sugar lowered you don't need stevia. The processed, powdered stevia is altered from its natural state. If you want sweet, use raw honey or maple syrup or sucanat or coconut sugar in moderation and at least you will be getting some nutritional value."

Nan

"I have tried all of the sugar alcohols and it's the same reaction – I get terrible, strong headaches for the whole day. It's the same with stevia, both the natural leaves and the drops. The headaches are so strong that I need to rest."

Gabriela

Addiction

Many people become addicted to stevia. As with drug addiction, the effect of stevia, its sweet taste, gets weaker, and a larger dose is needed to get the same "high" or sweetening effect. Also, the bitter aftertaste becomes less and less noticeable.

"My husband and I have been using pure stevia extract for at least 5 years. Lately, we've noticed it requires more to obtain a sweet taste. We also have a general loss of flavor in food. Checking the Internet, we found loss of taste is one side effect, so we quit using it for now. What we can't find is whether our taste buds will now recover."

Joan

"I have to admit that I consume a lot of stevia every day, I am finding it almost addicting, and seem to want to add more and more to my smoothies and tea. Can anyone help me out here, as to what to do, and how to quit drinking so much?"

Veronica

You know you are addicted to something when your desire to have it overcomes sound logic and common sense, and overpowers the best intentions. Switching from sugar to stevia does not cure sugar, or rather, sweet addiction, it only changes the form of the drug that fuels the fire of addiction.

ALLERGIES

Not everyone who uses stevia experiences adverse effects. The same is true with other non-caloric sweeteners. Some people may be more sensitive than others because they are allergic to stevia. The medical literature describes stevia allergy, which in some cases is so severe it can lead to life-threatening anaphylactic shock.

For instance, a seven-month-old girl had atopic eczema since she was two months old. She was treated with a steroid ointment, but it had no effect. She was breastfed. She had egg and cow's milk allergies, but no other known food allergies. Her mother cultivated stevia and drank herbal tea sweetened with stevia leaves. One day, her mother noticed that her daughter was chewing stevia leaves. She wasn't concerned because she considered stevia to be harmless. Minutes later the child was unconscious. The infant was rushed to the hospital where she was treated for anaphylactic shock and, fortunately, recovered. She was found to have an allergy to stevia leaves and steviol glycosides. Her mother stopped cultivating stevia and avoided eating anything with stevia. The infant had no further anaphylaxis and her atopic eczema cleared up. Even though the infant had eaten stevia only once, her mother had been consuming it daily. The mother's breast milk was causing an allergic reaction whose symptom was the infant's chronic eczema.

A two-year-old boy had atopic eczema since he was six months old. He had an allergy to eggs, but no other foods seemed to bother him. Medications had no affect on his eczema. One day, his mother gave him a drink of warm water sweetened with stevioside powder. Within minutes the child collapsed unconscious. He was rushed to the hospital and treated for anaphylactic shock, and recovered. He was found to have an allergy to stevia leaves and steviol glycosides. Thereafter, his mother stopped using stevia and removed it from the house. Her son had no further encounters with anaphylaxis and his longstanding eczema cleared up.

Allergist H. Kimata, MD, conducted a study of 200 infants, ages four months to two years, who visited the Morguchi-Keijinkai Hospital, in Japan, for allergy screening. Fifty infants were healthy, 50 had allergic rhinitis, 50 had bronchial asthma, and 50 had atopic eczema. None of the 50 healthy children were allergic to stevia. However, 16 percent of those with allergic rhinitis, 34 percent who had bronchial asthma, and 64 percent of the atopic eczema patients were found to be allergic to stevia leaf and stevioside. Dr. Kimata concluded that having preexisting allergies significantly increases the risk of being allergic to stevia.[2]

Stevia promoters acknowledge the fact that stevia allergy exists, but they claim such allergies are rare and that there is little need to worry. In fact, one independent research group, commissioned by Cargill to write a paper intended to ease people's fears that stevia may be an allergen, stated: "Some popular media reports and resources have issued food warnings alleging the potential for stevia allergy....Neither stevia manufacturers nor food allergy networks have reported significant numbers of any adverse events related to ingestion of stevia-based sweeteners....Therefore, there is little substantiated scientific evidence to support warning statements to consumers about allergy to highly purified stevia extracts."[3] But according to Dr. Kimata's study, stevia allergy may be far more common than suspected. Anaphylactic shock is a life-threatening event, and allergic eczema, asthma, and rhinitis are serious health issues for those that suffer from them. To claim that a stevia allergy is of no concern just so the sponsoring company's sales don't suffer seems callous and irresponsible.

Which symptoms are caused by a stevia allergy and which are caused by some aspect of stevia itself? It's hard to tell, but we can make a guess. Abdominal pain and cramping, nausea, and vomiting are most likely caused by the inability of digestive enzymes and intestinal bacteria to break down and digest steviol glycosides; the alteration of normal gut microbiota may also affect digestion and cause symptoms. Other low-calorie sweeteners that also do not digest well, such as sucralose and xylitol, are also known to cause digestive distress. Alterations in blood sugar and blood pressure could be the cause of the headaches and dizziness that some people report from using stevia.

Stevia is a member of the Asteraceae family of plants, which also includes ragweed, chrysanthemum, marigolds, and sunflower. Many plants from this family can induce allergic reactions in hypersensitive individuals. Some common allergy symptoms that might be associated with stevia include rhinitis (nasal irritation and inflammation), excessive mucus, cough, irritated throat, shortness of breath, muscle aches and pains, fatigue, skin rash (eczema), and canker sores.

Below are some comments from stevia users whose symptoms appear to have been allergy related:

"Saw some comments awhile back that stevia is in the ragweed family and a light bulb went on in my head. Maybe, just maybe, that's why I have had phlegm constantly in my throat the last couple years (about as long as I've been using it every day). So much so that one would think I was a smoker and I am not nor have I ever been. I stopped using it and the phlegm has subsided a lot!"
CJ

"Maybe this is why stevia always seems to bother me. I get really stuffed up and my glands swell. All sugar alcohols bother me as well. Severe bloating, gas, and diarrhea, makes me feel like my whole body is being squeezed. You can't get something for nothing."
Michelle

"Has anybody out there experienced throat irritation and coughing using stevia? I have had dry throat and coughing issues for over three years using this product, once in the morning for coffee and once at night with tea. I have seen numerous doctors and had all kinds of tests and they could not find anything wrong with my lungs or throat. I stopped using stevia three days ago and all symptoms are pretty much gone. I was concerned I had some exotic disease that nobody could diagnose. My father had ragweed allergies. I did not think I had any. Could this be the cause of my intense throat irritation? I thought I had throat or thyroid cancer. All negative"
Lynn

"I have a severe ragweed allergy and when I tried something with stevia in it (didn't realize it had stevia in it), it made me really sick (terrible throat irritation – felt like I'd swallowed a twig – and a horrible stomach ache). I wasn't sure what caused it so I didn't finish that food. A while later, I tried something else and had the same reaction...that was when I noticed that the box read "Made with Stevia!" I researched the last product I tried and saw that it, too, was made with stevia. I put that info together, called my allergist, and sure enough...ragweed family! Crazy!! I try to check all labels of new items, especially those labeled "All Natural!" prior to ingesting them but once in awhile something I ate/drank in the past updates its ingredients and adds stevia without me knowing. I can tell almost immediately when I try it (my lips start to tingle and my throat and stomach act wonky!) so I stop and add that to the list to not eat/drink. I've noticed that more and more companies are using stevia and that bothers me because while the reaction to stevia can be as severe as a peanut allergy, there isn't much info out there right now warning people."

Amy

"I usually avoid what I consider synthetics and any other elements I deem harmful or questionable. I recently got stevia for the first time, however, without knowing it. For a very long time I've used only Sierra Mist as a soda, because of the 'natural sugar' and lack of caffeine – but they changed the ingredients without saying so. My throat was on fire even as I drank a can of this 'new' drink. I feel certain it was the stevia. I'm gonna dump it all and be on the watch for stevia everywhere now."

Leon

CONTRAINDICATIONS

Using stevia if you are taking certain medications or have certain health conditions may cause adverse reactions. For example, stevia has a diuretic effect that may decrease how well the body gets rid of lithium, so if you are taking any medication that contains lithium, consuming stevia could cause dangerous side effects. Caution is also advised if you are taking diabetes medications that lower blood sugar. Many studies suggest that stevia can also lower blood sugar; therefore, if you are using stevia and taking diabetic medications to control blood sugar, the combination may lower blood sugar to dangerous levels. A similar situation exists if you take medication to lower your blood pressure; since stevia does that as well, the combination may lower blood pressure too much.

Caution has also been advised for certain health situations. If you have any food or pollen allergies, as noted above, your risk of being allergic to stevia is substantially increased.

Perhaps the biggest concern is for pregnant and nursing mothers. Some research suggests stevia may increase the risk of spontaneous abortion, or otherwise cause harm to the unborn baby. Even though this has not been firmly established, it is best to play it safe and avoid stevia during pregnancy. When breastfeeding, stevia compounds can and do find their way into mother's milk, exposing infants to all of stevia's potential health problems, which include allergies. Stevia should never be used by nursing mothers, or in infant formula or baby food.

Below is a partial list of medications that may cause adverse effects when combined with stevia.

Lithium Medications
Camcolit
Li-Liquid
Liskonum
Lithane
Lithicarb
Lithobid
Priadel
Quilonum

Diabetes Medications
Glimepiride (Amaryl)
Glyburide (DiaBeta, Glynase Pres Tab, Micronase)
Insulin
Pioglitazone (Actos)
Rosiglitazone (Avandia)
Chlorpropamide (Diabinese)

Glipizide (Glucotrol)
Tolbutamide (Orinase)

Blood Pressure Medications
captopril (Capoten)
enalapril (Vasotec)
losartan (Cozaar)
valsartan (Diovan)
diltiazem (Cardizem)
Amlodipine (Norvasc)
hydrochlorothiazide (HydroDiuril)
furosemide (Lasix)

SHARE YOUR EXPERIENCE

Some brands of stevia appear to cause more trouble than others. This may be due to the other ingredients added to the stevia. Truvia brand stevia appears to be one of the worst, with hundreds of adverse effects being reported.

If you or someone you know has experienced any adverse effects from stevia, I'd like to know about it. Write to me in care of the publisher of this book (see address on copyright page), or email me at info@piccadillybooks.com, and tell me what brand of stevia you used, your symptoms, and how you determined it was this product that was causing your problems. I would like to hear from you, please write and tell me your story.

8

Things You Probably Didn't Know About Stevia

MOST COMMERCIAL PRODUCTS ARE NOT STEVIA

You would think that stevia sweeteners would contain stevia (rebaudioside A, stevioside, or a mixture of steviol glycosides) as the only or at least the primary ingredient. But that isn't always the case. Most of the stevia sweeteners you buy in the store are not pure stevia, but a combination of ingredients of which stevia is only one. These other products can include artificial sweeteners, sugar alcohols, and sugars, which intensify the sweet taste and mask or dilute the bitter aftertaste of the stevia. For instance, Safeway brand Stevia Extract is primarily the sugar alcohol erythritol with some stevia added, along with preservatives and flavor enhancers. A brand called "Stevia in the Raw" sounds pure and natural, but the ingredient label reveals that it is primarily dextrose, not stevia (which is the second ingredient listed). Dextrose is a form of sugar usually derived from genetically modified (GMO) corn. Truvia brand stevia, which was developed jointly by Coca-Cola and Cargill, is a combination of erythritol (the main ingredient), rebaudioside A, and "natural flavors." PureVia brand stevia, which was developed by PepsiCo, contains in addition to rebaudioside A, dextrose, cellulose powder, and again, "natural flavors." Why a sweetener would need natural flavors is uncertain, and just what are these natural flavors anyway?

A common additive to many sugar substitutes, including stevia, is maltodextrin. It is used as a cheap bulking agent to bring the product to the approximate volume and texture of an equivalent amount of sugar. Maltodextrin is a polysaccharide—a form of sugar made of glucose. It is composed of a chain of 3 to 17 glucose molecules and is slightly sweet. Like other sugars it is quickly digested, with a glycemic index of 100, the same as glucose. It is ironic that a sugar that supplies calories and can raise blood sugar is combined with a zero-calorie sugar substitute.

Maltodextrin adds about four calories per packet of sweetener. The FDA allows any product containing fewer than five calories per serving to be labeled as having "zero" calories. In North America, maltodextrin is usually made from GMO corn.

I looked at a number of stevia products at my local store and found they contain a variety of other ingredients, including corn syrup solids (dehydrated corn syrup), citric acid, fumaric acid, tartaric acid, natural colors, sodium benzoate (a preservative), potassium sorbate (a preservative), erythritol made from corn (that's how it was listed on the ingredient label), artificial flavors, lactose (milk sugar), and isomaltulose (a form of sugar derived from sugar cane). If you are using stevia to avoid sugar and corn syrup, you had better read the ingredient label first.

Ingredient labels list the ingredients starting with the one that makes up the majority of the product, and going down in order to the one that makes up the least. In stevia sweeteners you would expect stevia to head the list, but it is often the second or third ingredient. Believe it or not, some stevia sweeteners contain almost no stevia at all, but you wouldn't necessarily know that from looking at the ingredient label.

Truvia, the most popular stevia product on the market, is not stevia at all. It is approximately 99 percent erythritol, with only a tiny amount of stevia. This fact was discovered accidentally by researchers in the Department of Biodiversity Earth and Environmental Science at Drexel University in Pennsylvania.

Interestingly, this discovery was initiated by a sixth grader. Few grade schoolers get their science project published in a major science journal, but that is what happened to 12-year-old Simon Kaschock-Marenda. Simon's father, Dr. Daniel Marenda, is a neurobiologist at Drexel University. Simon came to his dad with an idea for his sixth grade science fair project. He wanted to feed a variety of sugars and sugar substitutes to flies and see how the insects fared.

Simon and his dad purchased several sweeteners from their local supermarket for testing. One was Truvia. They mixed the sweeteners with food, put each in a container with fruit flies, and waited. By the end of the week, Simon pointed out that the flies in the Truvia container had all died, while the ones feeding on the other sweeteners were still alive. Thinking the results might have been a fluke, they repeated the experiment, only to obtain the same results. Flies raised on the Truvia sweetened food survived for only about six days, while those given table sugar lived for about 60 days. Realizing his son was onto something, Dr. Marenda moved the study from his home to his lab, called in other researchers to assist, and began a formal study.

The investigators fed groups of fruit flies one of each of the following: sucrose, corn syrup, Truvia, PureVia, Splenda (sucralose), Equal (aspartame and acesulfame K), and Sweet'N Low (saccharine). The fruit flies in all of the groups except one lived to their normal life expectancy in a lab of about 60 days. The flies fed Truvia died in only 5.8 days. The researchers wondered what was in Truvia that would cut the fruit flies' lifespan down by nearly 90 percent. They had tested PureVia, another stevia product, and the flies lived just about as long as those fed the other sweeteners, so stevia was eliminated as the toxic agent (see graph on page 104).

The first ingredient listed on the Truvia label is erythritol. So the investigators repeated the experiment using pure erythritol, and the result was almost identical to Truvia's. In fact, flies getting pure erythritol actually lived a day or two longer than those fed Truvia. (see graph on page 105). The investigators discovered that Truvia was almost entirely erythritol.[1]

Truvia is twice as sweet as sugar. However, rebiana (rebaudioside A), the form of stevia in Truvia, is 200 times sweeter than sugar. Erythritol is about the same sweetness as sugar. Do the math. How much erythritol would you have to mix with rebaudioside A to have a product that is twice as sweet as sugar? The answer: 99 percent erythritol and 1 percent rebaudioside A. To call Truvia a stevia sweetener is deceptive advertising.

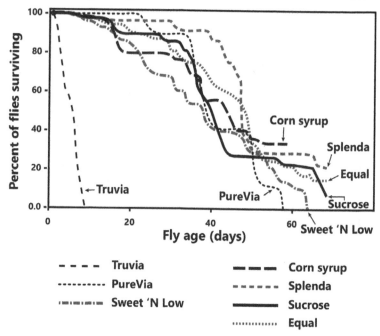

Comparing non-nutritive sweetener effects

Percent of flies surviving vs. Fly age (days)

Corn syrup
Splenda
Equal
Truvia
PureVia
Sucrose
Sweet 'N Low

– – – – Truvia	▬ ▬ ▬ Corn syrup
·········· PureVia	– – – – – Splenda
·—·—·—· Sweet 'N Low	▬▬▬ Sucrose
············ Equal	

The reason the makers of Truvia call it stevia is obvious; to cash in on stevia's wholesome image as a harmless herb. Also, erythritol is cheaper to produce and doesn't have a bitter aftertaste, which is why Truvia has no aftertaste—another sign that it really isn't stevia. In fact, any brand of stevia that lacks the bitter or metallic aftertaste probably has very little stevia in it. Many companies purchase their stevia from Cargill and repackage it under their own label, with no mention of the trade name Truvia. Look at the ingredient label of any brand of stevia and if you see the combination "Erythritol, stevia, and natural flavors," in that order, it is probably Truvia.

Now, you might assume that erythritol isn't so bad. It is a sugar alcohol that has the sweetness of sucrose but fewer calories. Because of this, however, it has many of the same faults as other low-calorie sweeteners and then some. Since it is not well absorbed in the digestive tract, it can cause digestive disturbances, gas, bloating, and diarrhea. The erythritol used in Truvia is derived from a yeast

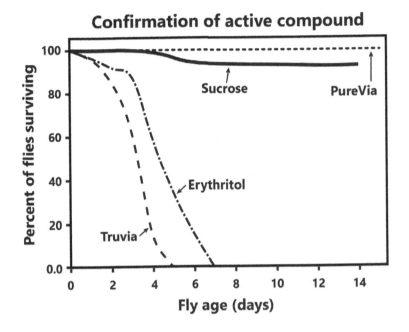

Confirmation of active compound

Y-axis: Percent of flies surviving (0.0 to 100)
X-axis: Fly age (days) (0 to 14)

Sucrose

PureVia

Erythritol

Truvia

organism fed the GMO corn maltodextrin, which negates Cargill's claim that Truvia is "natural." One good thing about Truvia, is that it makes a good insecticide. That was the conclusion that the Drexel University researchers came to, and they recommended Truvia as such. So if you happen to have an infestation of insects in your gut, Truvia might be a possible remedy.

STEVIA IS AN ARTIFICIAL SWEETENER

It doesn't really matter what you call them—nonnutritive, zero-calorie, non-caloric, low-calorie, or artificial—they are all terms used to describe sugar substitutes that provide the sweet taste of sugar without the calories.

Marketers like to call stevia a "natural" or "herbal" sweetener to distinguish it from the competition and project a healthy or wholesome image. But in reality, stevia is no different from any other artificial sweetener. The argument that it is not artificial because it comes from a natural source can also be made for aspartame, which is composed of two amino acids (aspartate and phenylalanine) and a little bit of alcohol—all of which occur naturally.

Likewise sucralose. It is a combination of sugar and chlorine. Sugar occurs naturally, as does chlorine. (Ordinary table salt consists of one molecule of chlorine and one of sodium; chemists call it sodium chloride.) One might object that sugar is a highly refined product. But so is stevia: stevia sweeteners consist of steviol glycosides that are highly processed, refined, and purified, just like sugar and chlorine.

Indeed, stevia is no more natural than any other artificial sweetener. You never find stevioside or rebaudioside A occurring by themselves in nature; they don't exist that way. Therefore, the stevia sweeteners consisting of purified steviol glycosides are man-made, just like other artificial sweeteners. Even the FDA states that steviol glycosides "are not stevia," and that stevia sweetener is not an herb. It is a purified chemical.

Artificial sweeteners are very different from one another chemically, but what ties them together is their effect on our bodies. The four major characteristics that all artificial sweeteners have in common are:

1) Does not occur in nature in its present form

2) High-intensity sweet taste, generally over 100 times sweeter than sucrose (table sugar)

3) Provides no significant calories

4) Is not immediately or noticeably harmful, but with repeated use causes changes in glucose metabolism, promotes insulin resistance and weight gain, slows metabolism, initiates inflammation, alters gut microbiota and the intestinal environment, and blocks the formation of ketone bodies—all of which can lead to chronic degenerative disease.

Since stevia shares all these characteristics with all other artificial sweeteners, it should be classified as one. It definitely shouldn't be classified as an herb or natural product. We should stop referring to it as an herbal sweetener and start calling it what it is, an artificial sweetener.

A BITTER AFTERTASTE

Another characteristic that stevia shares with some, but not all, synthetic sweeteners is a bitter aftertaste. Stevia is well-known for its bitter aftertaste, and many people don't like to use it for that reason.[2] If a brand of stevia sweetener does not have the characteristic aftertaste, it's not because the manufacturer has some secret processing technique, or that the stevia was grown in special organic soils (as I've heard some promoters claim); it means that it has been heavily diluted with other sweeteners and fillers, and is not pure steviol glycosides. Acesulfame K, cyclamate, and saccharin also have a characteristic bitter aftertaste, and for this reason are often combined with other sweeteners. Combining sweeteners enhances the sweetness while diluting the bitter aftertaste.

Most toxic compounds in nature are bitter. Scientists have long believed that a bitter taste evolved as a defense mechanism to detect poisonous compounds in plants and bacterial toxins in spoiled foods. There is good scientific evidence to support this hypothesis.[3] Although there are bitter compounds in nature that are not particularly dangerous, our natural instinct when we eat something bitter is to spit it out or stop eating it. If we happen to swallow it, the taste receptors in our GI tract sense it and can initiate a series of reactions that protect us from harm, such as triggering an immune response or initializing regurgitation, diarrhea, or digestive discomfort. This may be one of the reasons why artificial sweeteners often cause digestive distress.

We also have taste receptors in our sinuses that can detect bitter secretions from potentially harmful bacteria. Activation of the bitter taste sensors in our sinuses triggers a defensive immune response against inhaled pathogenic organisms.[4]

The bitter taste in our mouths is a natural universal warning sign to stop eating. The vary presence of the aftertaste associated with stevia is a warning that we should not be eating it.

WHERE DOES STEVIA COME FROM?

If you believe the story that stevia comes from the jungles of Paraguay, you are in for a surprise. Many brands of stevia are promoted as originating in Paraguay or Brazil, portraying an image that the product comes straight from deep within South American forests, like some exotic medicinal herb that is harvested in the wild by local villagers. Nothing could be farther from the truth.

Stevia is grown commercially in Central America, Korea, Paraguay, Brazil, Thailand, and, most notably, China. Most of it is grown on large commercial corporate farms and processed in modern factories that refine and purify the extract down into its pure crystalline steviol glycosides. About 85 percent of all the stevia sweeteners you see in the store and on the Internet come from China. The world's top stevia sellers (e.g., Cargill, Merisant, etc.) import all their stevia from China. Most of the health food store brands, such as Now and KAL, also come out of China. Even if a brand claims its stevia is *grown* in Paraguay, it is then usually shipped to China for processing. You're not usually told the latter part.

The grossly underpaid (should I even say slave) labor force, cheap (and questionable) extracting agents used, and flagrant disregard of the environment and massive pollution, lower the cost, making manufacturing stevia cheaper in China, even cheaper than it can be produced in South America. Food and supplement companies flock to China for their dirt cheap products. This wouldn't be such a big issue if the workers received a fair wage, and if the manufacturers didn't often substitute toxic chemicals in the processing, or adulterate the products with cheaper filler materials, as they are often known to do.

This is a bit scary, seeing that China has a horrible record when it comes to product purity and safety. I've seen too many horror stories in the news about food and drug products coming from China that have been adulterated with cheap and sometimes deadly additives (not listed in the ingredient label) or processed with toxic chemicals that leave residues in the products that have led to serious health problems.

One well publicized example is the Menu Foods fiasco of 2007. Commercial pet foods are required to have a certain amount of protein and other nutrients to meet the standards of the American Association of Food Control Organization. When checking compliance of food products, the protein content is not measured directly but is calculated based on nitrogen content. Protein is roughly 16 percent nitrogen by weight, so the nitrogen levels in pet foods are measured to determine (or estimate) the quantity of protein present. Unfortunately, there are other substances that contain nitrogen that can mimic protein.

Menu Foods, a company that produces dog food for most of the name brand pet food companies, imported wheat protein (gluten) from China that was tainted with the chemical melamine. Melamine contains 67 percent nitrogen. Visually, wheat flour is indistinguishable from wheat gluten, and one could easily be mistaken for the other. The Chinese supplier mixed inexpensive and low-protein wheat flour with melamine to produce a nitrogen reading consistent with that of gluten. A nitrogen analysis would not have shown anything wrong. Wheat flour is much cheaper than wheat gluten. If it weren't for one oversight by the Chinese supplier, nobody would have been the wiser. The problem was that melamine is poisonous. The adulterated wheat flour was used in the manufacture of hundreds of pet food products. These tainted pet foods were then sold across the country, resulting in massive illness and numerous deaths of dogs and cats. Over 260 dog and cat food products were recalled, including foods for horses, fish, and reptiles. If it weren't for the deaths, probably no one would have known that these tainted pet foods were protein deficient, and they would have continued to be sold to unsuspecting pet owners for years. The lack of protein would eventually contribute to protein deficiency in the animals and over time they would become sick and die. Cases like this make you wonder what other ingredients or products have come out of China that have been made with cheaper and potentially toxic ingredients?

In August 2008 executives of the New Zealand company Fonterra Group, the world's largest exporter and importer of dairy products, met with business partners at the headquarters of Sanlu Group, China's largest powdered milk producer. The Fonterra executives were deeply troubled. They had recently discovered melamine, the same chemical that sickened and killed pets throughout the US, in their powdered infant milk formula.

Government testing over the next few weeks discovered that the problem was not limited to Sanlu. The products of at least 20 dairy companies around the country were found to be contaminated with melamine. Most of the blame was placed on the dairy farmers. To keep costs low, they were selling watered-down milk with melamine added to boost the tested protein level. The dairymen, in turn, blamed the operators of the thousands of milk collection stations scattered across the country, which purchase raw milk with little regulatory oversight.

It was reported that 54,000 infants and children in China became ill after drinking the contaminated milk, some of whom died. To make matters worse, the chemical was even found in China's most famous candies, White Rabbit, which is also sold abroad. Melamine was also discovered in the products of international companies, including giants such as Cadbury, Nestle, and Unilever.

When the Fonterra Group met with their Chinese partners, it was just days before the start of the Beijing Olympics, and Chinese authorities were hypersensitive about anything that might stain the country's image. Over the next five weeks Fonterra engaged in a nerve-racking battle with their Chinese partner over what to do. Fonterra wanted to break the news and stop the sale and distribution of the tainted product, but the Chinese government would not allow it, at least not until the Olympics were over. So they waited. In the meantime, thousands of children were sickened and hospitalized and tainted products were shipped abroad. Lives didn't matter, the country had to uphold its image. The government sent out an order that nothing was to negatively affect the Olympics. The executives from Fonterra and Sanlu were sternly threatened to keep quiet.

Finally, Fonterra sought help from the New Zealand government to try to work things out diplomatically. Two weeks after the Olympics ended, and a little over five weeks after Fonterra first brought up their concerns, the Chinese government suddenly jumped into action as if it were a new discovery. They ordered a recall of the products and arrested the dairy farmers, the Sanlu officials, and anyone else who might have been involved, all as a show of concern for the public's safety.

As soon as the news broke, the Chinese government ordered the media to use only reports generated from official government news organizations. Stories on websites that had already reported the event were erased. A dairy company that was not involved in the scandal said they were all ordered by the government not to speak with anyone from the media.

It seems China is rampant with corruption and fraud. Steamed buns stuffed with meat are a popular Chinese food. The buns are stuffed with pork, but meat is expensive and cardboard is not. Unscrupulous manufacturers soak cardboard in chemicals to make them soft, then mix it with pork fat and spices and stuff it into the buns—yum! It's bad enough to eat cardboard, but how safe are the chemicals it's soaked in?

Rats are plentiful in China, so why not eat them? They do, without telling the customer. Rat meat is passed off as beef. Reports say it is common to use banned chemicals to process the meat and inject the meat with water to increase its weight.

In 2010, as many as 50 factories in southern China were discovered manufacturing rice noodles made from rotten grain and possibly poisonous flavorings. These factories are believed to have made up to 1.1 million pounds of shoddy noodles every day before being exposed.

Rice is a mainstay in China. The people eat it daily. If rice isn't cheap enough already, they now have fake rice made from a mixture of potatoes and plastic. It is said that eating three bowls of this fake rice would be the equivalent of eating one plastic bag. The plastic rice looks very much like natural rice so it is not easy to tell them apart. One source reported that cooking this fake rice in soup will form a thin plastic film over the top. It has been reported that plastic rice is being exported to other Asian countries such as Vietnam, Singapore, Indonesia, and India.

While most of these adulterated foods were consumed in China, who knows how many of them have been exported elsewhere?

China is the counterfeit king of the world. Any product that is manufactured in North America or Europe, that has any sales potential at all, is almost guaranteed to be copied. Fake Splenda is counterfeited and sold abroad. Who knows what really is in these counterfeit Splenda products. The same thing could be happening with stevia.

What about organic certification? Doesn't that guarantee a better product? Not at all. Even the organic brands of stevia are suspect. I've seen farms in Asia that claim to be organic, and have a certificate to prove it, yet they didn't get the certificate from an organic certifier, they paid for a copy. All it takes is one company to get organic certification, then they make copies and sell it to other farms and factories so they, too, can claim organic status. This is a common practice in some Asian countries. After all, if they are going to put chemicals like melamine in foods, what's going to stop them from putting non-organic ingredients in organic labeled products?

9

Artificial Sweeteners

SACCHARIN

Saccharin is the granddaddy of all artificial sweeteners, having been discovered by accident in 1879 by Constantin Fahlberg, a German-American chemist working on coal tar derivatives at Johns Hopkins University. Fahlberg was searching for a food preservative when he stumbled upon the highly sweet compound.

One evening, Fahlberg was so involved in his laboratory work that he forgot to go for supper until late. When he realized the time, he rushed off for his meal without stopping to wash his hands. He sat down to eat, broke a piece of bread, and put it to his lips. It tasted unspeakably sweet. He rinsed his mouth with water and dried his moustache with a napkin, when, to his surprise the napkin tasted sweeter than the bread. He was puzzled. He took a drink of water, and the goblet had a syrupy sweet taste where he had touched it. He tasted the tip of his thumb and found it surpassed the sweetness of any candy he had ever eaten. Realizing the sweetness must have come from one of the coal tar derivatives he was working on in his lab, he sprang from the dinner table and ran back to the laboratory. There in his excitement, he tasted the contents of every beaker on the table until he found the one with the sweet taste. Fortunately, none of the beakers contained any poisonous liquid.

In 1884, working on his own in New York City, Fahlberg applied for patents in several countries, and named the substance saccharin after the term saccharide, which is derived from the Greek word for sugar. Two years later, he began production of the new sweetener in a suburb of Magdehurg, Germany.

Saccharin is heat stable and does not react chemically with other food ingredients, making it well suited as an ingredient in prepared foods and beverages. Saccharin passes though the digestive system without being digested or providing any calories. It is about 300 times sweeter than sucrose (table sugar), with a bitter or metallic aftertaste, especially at high concentrations. To reduce the bitter aftertaste, saccharin is often combined with other sweeteners. A 10:1 cyclamate-to-saccharin blend is common in countries where both sweeteners are permitted. This blend results in more sweetening power with less of the off taste associated with each. Saccharin is often used with aspartame in diet fountain drinks, allowing the beverage to maintain sweetness should the fountain syrup be stored beyond aspartame's relatively short shelf life.

At the time saccharin was introduced it was used as a cheap replacement for sugar in food manufacturing, but was not well accepted by people in general; many did not like the bitter aftertaste. Some scientists expressed caution in using the artificial sweetener, as it had not been adequately tested for safety. In 1907, Harvey Wiley, the director of the bureau of chemistry for the FDA, viewed it as inferior to sugar and wanted it banned, at least until further studies could be done to prove its safety. However, with sugar rationing during World War I and again in World War II, saccharin sales took off.

Animal studies in 1948 to 1949 by scientists at the FDA raised the first warning that saccharin might cause cancer and kidney disease.[1] In 1958, the United States Congress amended the Food, Drug, and Cosmetic Act of 1938 with the Delaney clause to mandate that the FDA not approve substances that induce cancer in humans or animals. Studies published in the early 1970s linked saccharin with bladder cancer in rats and mice.[2] As a consequence, Canada and several other countries completely banned the sweetener. The FDA proposed a ban in 1977. But by then saccharin had been used by dieters and diabetics for many years. The Calorie Control Council, which represents the diet food and drink industry, ran an ad campaign encouraging consumers to protest the ban. The campaign

114

succeeded in inciting a public outcry against the ban because there was no other sugar substitute available at that time. Due to public pressure, Congress stepped in and placed a moratorium on the ban and mandated further study of the sweetener's safety. In the meantime, saccharin was allowed to remain on the market as long as the label contained the following warning: "Use of this product may be hazardous to your health. This product contains saccharin which has been determined to cause cancer in laboratory animals."

For the next 20 years the food industry sought to eliminate the cancer warning label. A number of studies were published that failed to find any clear link between saccharin and cancer in humans and animals other than male rodents. The reason given why only male rats were susceptible was that they have a unique combination of high pH, high calcium phosphate, and high protein levels in their urine. One or more of the proteins prevalent in male rats combine with the calcium phosphate and saccharin to produce sharp microcrystals that damage the lining of the bladder. Over time, the rat's bladder responds by overproducing cells to repair the damage, which leads to tumor formation. Since this does not occur in humans, it was claimed that we were at no risk of developing bladder cancer.

The Calorie Control Council petitioned the government to repeal the warning label and remove saccharin from its list of cancer causing chemicals. A number of scientists opposed the lifting of the warning, and urged the government to keep the sweetener on its list of cancer causing chemicals. Despite their protests, saccharin was delisted as a cancer risk. In a joint letter, a group of scientists told the National Toxicology Program (NTP), a division of the National Institute of Environmental Health Sciences, that declaring saccharin safe would "result in great exposure to this probable carcinogen in tens of millions of people, including children (indeed, fetuses). If saccharin is even a weak carcinogen, this unnecessary additive would pose an intolerable risk to the public."

Samuel Epstein, MD, a professor of environmental medicine at the School of Public Health, University of Illinois Medical Center, and a co-signer of the letter to the NTP said, "In light of the many animal and human studies clearly demonstrating that saccharin is carcinogenic, it is astonishing that the NTP is even considering delisting saccharin."

In their letter to the NTP, the scientists described several rodent studies that showed saccharin caused cancer in the bladder, uterus, skin, lungs, and other organs, and that it promoted chronic kidney disease. Bladder tumors were found in both male and female rats, discrediting the strongest argument for saccharin's presumed safety, i.e., that it is just male rats that are susceptible. In addition, they cited six human studies, including a large National Cancer Institute study, that found an association between bladder cancer and heavy saccharin use.[3]

Despite opposition from scientists, the warning label in the US was repealed on December 21, 2000. Soon thereafter, many countries that had banned saccharin lifted their bans as well. To date, saccharin has not been proven completely safe. Just because some studies show no link to cancer does not negate those studies that do.

Most safety studies are not really efforts to determine if a sweetener is safe, but to see if it is toxic or carcinogenic. If the study finds no acute toxicity, then the sweetener is declared safe. But a lack of acute toxicity does not mean safe or healthy. Sugar is considered nontoxic, but heavy long-term use can lead to a host of health problems. The same could be true for artificial sweeteners as well. Long-term studies are rare and are generally focused on determining carcinogenicity; however, more subtle health consequences are ignored or overlooked, such as liver, kidney, brain or spleen dysfunction which may not be readily noticeable. And if an investigator is not looking closely for a specific problem, it may go undetected.

CYCLAMATE

Cyclamate is the second oldest of the artificial sweeteners. It was discovered in 1937 by Michael Sveda, a graduate student working on his PhD at the University of Illinois. Sveda was in the lab working on the synthesis of anti-fever medication. He put his cigarette down on the lab bench, and when he put it back in his mouth, he discovered a sweet taste. He quickly realized the sweetness was from one of the chemical solutions he was working on. He tasted the contents from each beaker in front of him to find the one with the sweet taste. At the time, the only sugar substitute available was saccharin, but it had a strong aftertaste and the market was ready for a better sweetener. Sveda eventually applied for a patent for the new sweetener he called sodium cyclamate.

The patent was purchased by DuPont and later sold to Abbott Laboratories, which undertook the necessary studies for FDA approval. Abbott intended to use cyclamate to mask the bitterness of certain drugs such as antibiotics and pentobarbital. In 1958, it was given GRAS (Generally Recognized As Safe) status by the FDA. By this time obesity was becoming an issue, and Abbott began to advertise cyclamate as a low-calorie sweetener rather than as a cheaper alternative to sugar.

Cyclamate is only 30 times sweeter than sugar, far less than saccharin, and is the least sweet of all the artificial sweeteners. Cyclamate was marketed in tablet form as well as a liquid for use by diabetics as an alternative tabletop sweetener. Cyclamate is heat stable, making it suitable for use in cooking and baking.

Like saccharin, cyclamate has a bitter aftertaste. However, mixing 10 parts cyclamate with one part saccharin yields a product with little aftertaste. The product was sold as Sweet'N Low. By the late 1960s, large amounts of cyclamate were being consumed by the American public in products ranging from soft drinks to salad dressings.

Cyclamate's downfall started in 1966 when a study reported that intestinal bacteria could convert the sweetener to cyclohexylamine, a toxic compound. Cyclohexylamine is toxic when ingested or inhaled, and can be fatal. It is used to make the herbicide hexazinone. This discovery triggered a series of rat feeding studies to determine its health risks. A 1969 study found the common 10:1 cyclamate to saccharin mixture to cause bladder cancer in rats. In 1970 the FDA revoked the GRAS designation from cyclamate and banned its use from all food and drug products in the United States.

The Canadian and UK governments followed, but instead of a total ban limited its use to that of a tabletop sweetener only. Other countries followed by banning the product or restricting its use. Sweet'N-Low and Sugar Twin sweeteners sold in the United States are saccharin-based, but in Canada, where saccharin was banned until 2014, the versions of each were cyclamate-based.

As with saccharin, the food industry tried to exonerate cyclamate from any wrongdoing by sponsoring studies to prove it harmless or at least non-carcinogenic. So far, they have not been successful. A number of studies have been published, but none of them have been convincing. For example, in 2000 a paper was published describing the results of a 24-year-long experiment in which 16 monkeys were fed a normal diet and 21 monkeys were fed either 100 or 500 mg/kg cyclamate per day; the lower dose corresponds to the amount of sweetener in about six cans of a diet beverage, and the higher dose corresponds to about 30 cans. Two of the higher-dose monkeys and one of the lower-dose monkeys were found to have malignant cancer, each with a different kind of cancer; in addition, three benign tumors were found. The authors concluded that cyclamate was not carcinogenic because the cancers were all different and there was no way to link cyclamate to each of them; this interpretation of the results has been contested. If cyclamate causes cancer, even in just a few subjects, it doesn't matter what type of cancers they are.

Cyclamate has no advantage over other artificial sweeteners, and is far less sweet. Since the patent protection for competing products—saccharin, aspartame, and acesulfame K—have all expired, allowing any company with the resources to produce and sell these products, there is little motivation to push to remove the ban on cyclamate. It is not likely that the restrictions on cyclamate will be lifted anytime soon.

ACESULFAME POTASSIUM

Acesulfame Potassium is listed on food ingredient labels as acesulfame potassium, acesulfame K, or Ace K (K being the chemical symbol for potassium), and is marketed under the trade names Sunett and Sweet One. After ingestion, acesulfame K is completely absorbed and then rapidly eliminated in the urine without providing any calories. It is 200 times sweeter than sucrose and just as sweet as aspartame. Like several other artificial sweeteners it has a bitter aftertaste. It is often blended with sucralose or aspartame to dilute the aftertaste. Unlike aspartame, acesulfame K is heat stable and so does not lose it sweetness when heated, allowing it to be used as a food additive in baking and in packaged, prepared food products.

Like saccharin and cyclamate, acesulfame K was discovered accidentally. The discovery was made in 1967 when Karl Clauss, a chemist working for Hoechst Chemical Company in Germany, licked his finger to pick up a piece of paper and noticed a sweet taste. He realized the sweetness was from residue of the chemical he was working with. He immediately recognized the commercial potential of his finding and spent the next several years doing the necessary research and testing to get the product approved as a sugar substitute.

Of all the artificial sweeteners, acesulfame K has received the least rigorous safety research. The tests carried out by Hoechst Chemical Company in the 1970s have been criticized as poorly designed and conducted. Not enough mice were used in the studies to have any significance, the dosages were too low, and the durations of the studies were far too short to determine any real measure of safety. If and when tumors appeared, they were simply ignored as not being relevant.

To compound the problem, acesulfame K was often contaminated with methylene chloride, a carcinogenic solvent used in cleaning the production materials. Many of the early studies linked the sweetener to multiple cancers in laboratory animals, which may or may not have been caused or exacerbated by the methylene chloride.

Even without the methylene chloride contamination there has been concern about the sweetener's carcinogenic potential. While short-term studies have shown no indication of carcinogenicity, some long-term studies have.[4] These latter studies are generally dismissed because the dosages are far greater than any human would consume.

Despite these concerns, acesulfame K was approved for use in Europe in 1983. In the United States, the FDA approved it for limited use in 1988, for use in soft drinks in 1998, and for general use in 2003.

Since then, further research has identified new concerns. Although acesulfame K has a stable shelf life, it can degrade to acetoacetamide, which is of concern because it is toxic. A 2008 study found acesulfame K causes DNA damage.[5] Another potential problem is that acesulfame K can be passed through the placenta and mammary glands to offspring.[6] A study of 20 lactating women, 14 of whom reported using artificial sweeteners, found that acesulfame K was the most commonly found artificial sweetener in breast milk. The breast milk of 13 of the women, including some who reported no intake of artificial sweeteners, contained acesulfame K. Apparently, the women who did not purposely use artificial sweeteners were still exposed to them in as ingredients in packaged foods. How acesulfame K affects infants is unknown, but it would be wise for pregnant and nursing women to make special efforts to avoid this and other artificial sweeteners.

While carcinogenicity seems to be a major focus with studies on artificial sweeteners, it certainly is not the only concern. Most studies are too brief to demonstrate the effects of long-term usage. People who use artificial sweeteners are expected to use them for life, over a period of many years, but long-term studies are few. One 40-week study showed a moderate effect on neurometabolic function, suggesting chronic usage of acesulfame K may alter brain function and impair learning ability.[7] The amount of acesulfame K given to the lab animals was equivalent to that which would typically be consumed by a human. In the treated mice, it took only 10 months to show alterations in brain function resulting in reduced memory and learning ability as compared to untreated mice. What effect, then, would it have on humans consuming it daily for 10, 20, or more years? And what effect would it have on the brain of a fetus or nursing infant of a mother consuming the sweetener?

ASPARTAME

Aspartame was first synthesized in 1965 by James M. Schlatter, a chemist working for G.D. Searle & Company. Like the artificial sweeteners before it, aspartame was discovered by accident, in this case while trying to develop a new anti-ulcer drug.

Aspartame was approved by the FDA as a tabletop sweetener in 1981 and as a general purpose sweetener in 1996. Several countries in Europe approved aspartame in the 1980s, with the full European Union approval in 1994.

In 1985, Monsanto Company bought G.D. Searle, and aspartame became a separate Monsanto subsidiary called the NutraSweet Company. In 2000, J.W. Childs Equity Partners purchased the NutraSweet Company from Monsanto. The European patent on aspartame expired in 1987 and the US patent in 1992. It is currently sold under the brand names NutraSweet, Equal, Spoonful, and Aminosweet.

Aspartame is about 200 times sweeter than sugar. Unlike previous synthetic sweeteners, aspartame is completely broken down by the body into its individual components—amino acids (aspartic acid and phenylalanine), and methanol (wood alcohol). Even though it delivers about 4 calories per gram when digested, the quantity of aspartame needed to produce a sweet taste is so small that its caloric content is negligible.

Aspartame is the most controversial of all the artificial sweeteners. Since its approval, it has accounted for over 75 percent of the adverse reactions to food additives reported to the FDA. At least 90 different symptoms have been documented as being caused by aspartame. Some of these include: headaches/migraines, dizziness, seizures, nausea, numbness, muscle spasms, rashes, depression, fatigue, irritability, insomnia, vision problems, hearing loss, heart palpitations, breathing difficulties, anxiety attacks, slurred speech, loss of taste, tinnitus, vertigo, memory loss, joint pain, cancer, and weight gain. In addition, aspartame may trigger or worsen brain tumors, multiple sclerosis, epilepsy, chronic fatigue syndrome, Parkinson's disease, Alzheimer's disease, fibromyalgia, and diabetes.

Aspartame breaks down into its constituent amino acids under elevated temperature or high pH (alkaline). This makes aspartame unsuitable for baking or cooking. Despite this limitation, the sweetener has found its way into about 6,000 products. Most soft drinks have a pH between 3 and 5, slightly acidic, where aspartame is reasonably stable. At pH 7 (neutral), it degrades very quickly and can lose its sweetness in a matter of days. In syrups used for fountain drinks, aspartame is often blended with a more stable sweetener, such as saccharin.

The three breakdown products of aspartame are all toxic in high doses. When the temperature of aspartame exceeds 86 degrees

F (30° C), the methanol in aspartame converts to formaldehyde and then to formic acid, which in turn causes metabolic acidosis. The methanol toxicity mimics multiple sclerosis; thus, people may be misdiagnosed with having MS.

Phenylalanine is an essential amino acid which must be included in the diet, but sustained high blood levels can lead to brain damage. This is of major concern to those who are born with an inherited condition called phenylketonuria or PKU. These people cannot metabolize phenylalanine, which then builds up to dangerous levels in their brains. This means that aspartame, due to its phenylalanine content, is not suitable for PKU sufferers and consequently requires a warning to that effect on products in which it is an ingredient. It has been suggested that some of the side effects of aspartame use may be caused by a sudden increase in brain phenylalanine levels, whether or not the person has PKU. Risk is especially high when the sweetener is consumed along with foods high in carbohydrates. Carbohydrates trigger insulin release into the bloodstream which, in turn, makes it easier for phenylalanine to cross the blood-brain barrier.

Aspartic acid, another amino acid, can also be toxic to the brain in high doses. At high doses aspartic acid becomes an excitotoxin, and has been shown to cause brain damage in animals.[8]

In light of continued research that is raising questions about aspartame's safety, many scientists are calling for government agencies to reconsider regulations governing aspartame's widespread use in order to better protect public health. For instance, Italian researchers at the Cesare Maltoni Cancer Research Center added aspartame to the standard diet of rats, using dosages designed to simulate a wide range of human intakes. Each animal was observed from eight weeks of age until death. This is in contrast with earlier studies that typically followed animals for only 110 weeks of age or less, corresponding to only two-thirds of a rat's lifespan (in humans, approximately 80 percent of cancer diagnoses are made in the last third of life, after age 55). Deceased rats were examined for microscopic changes in various organs and tissues, enabling a comprehensive assessment of aspartame's carcinogenic potential. The study involved a total of 1,800 animals, far more than in previous studies, allowing for highly statistically significant results.

Aspartame-fed rats showed significant evidence of lymph-omas/leukemias, carcinomas of the renal pelvis and ureter, and other tumorous growths.[9] These effects were evident at daily doses that were equivalent to less than half the recommended daily dosage considered safe for humans, thus indicating that even at relatively low doses aspartame is potentially dangerous.

As with all other artificial sweeteners, as soon as studies or consumer complaints arose suggesting that the sweetener causes harm, the makers quickly responded by sponsoring bogus safety studies to ease consumers' fears and confuse the medical community. The results of studies that suggest problems with aspartame are explained away. For example, the Italian study mentioned above was brushed aside with the excuse that the multiple cancers that occurred were caused simply by chance. The negative publicity associated with aspartame has been so substantial that the makers have sponsored studies not only to demonstrate its safety but to create the illusion that it is even healthy. Studies have been published that allegedly show that aspartame possesses antipyretic, analgesic, and anti-inflammatory actions, and that it aids in relief of some chronic conditions such as arthritis. In one study, which involved diabetics, subjects in the placebo group experienced more adverse reactions than those in the aspartame group, demonstrating that aspartame is not only harmless but even healthful! Everyone should be taking it to improve their health—a total absurdity, but that is what the study was essentially suggesting. Unfortunately, so many of these "marketing" studies are published that the truth is drowned out and the sweetener remains on the market.

Ralph G. Walton, MD, chairman of the Center for Behavioral Medicine at Northeastern Ohio Universities College of Medicine, surveyed the peer reviewed medical literature for safety studies on aspartame. He identified 166 studies that were published between 1970 and 1998 that had relevance concerning human safety. Out of that number, 74 were sponsored by the artificial sweetener industry and 92 were independently funded. Every one (100%) of the industry funded studies attested to aspartame's safety, whereas 92 percent of the independently funded research identified safety concerns with the sweetener.[10]

The FDA and advisory committees of the European Food Safety Authority have reviewed the studies on aspartame and have concluded that there is still not enough substantive evidence that aspartame is harmful to the general public, although they do acknowledge possible problems for select individuals, such as those with PKU, allergies, or susceptibilities to certain conditions such as seizures. It seems like anyone who suffers from any adverse reaction from aspartame must, in their opinion, be among the unlucky few who are "susceptible."

SUCRALOSE

Sucralose was discovered in 1976 by British researchers working for Tate & Lyle and Queen Elizabeth College (now part of King's College London). While researching ways to use sucrose and its synthetic derivatives for industrial use, one of the scientists was told to "test" a chlorinated sugar compound. He thought he was asked to "taste" it, so he did. He found the compound to be exceptionally sweet.

Tate & Lyle patented the sweetener in 1976. Sucralose was first approved for use in Canada in 1991, in Australia in 1993, in New Zealand in 1996, in the United States in 1998, and in the European Union in 2004. Patent protection on sucralose has expired, so the product is sold under a variety of brand names including Splenda, Zerocal, Sukrana, SucraPlus, Cukren, and Nevella.

The sweetener is produced by combining chlorine atoms with sucrose. The resulting compound has a sweetness that is about 600 times greater than sucrose. The original sugar molecule is altered into a form that does not occur in nature and, therefore, our bodies do not have the ability to properly metabolize it. As a result, only about 15 percent is absorbed in the digestive tract and metabolized, with the rest being excreted in the urine and feces. Since these molecules are no longer sugar, but altered versions of sugar, we don't really know what effect they can have on the body and on our health. Because such a small amount is needed to sweeten foods, it provides essentially no calories. Sucralose is heat stable, meaning that it stays sweet even when used at normal cooking temperatures, and therefore can be used as a sugar substitute in baked goods.

Sucralose has some definite advantages over other artificial sweeteners. In addition to being heat-stable, unlike some, it does not have a bitter aftertaste, stores well, is twice as sweet as saccharin and three times as sweet as aspartame and acesulfame K, is relatively inexpensive, and has not yet had the stigma associated with other artificial sweeteners, especially aspartame.

Sucralose has been promoted as a more "natural" sweetener than other nonnutritive sweeteners because it is made from sugar. It is supposedly healthier than sugar because it contains no calories, and therefore may aid in weight loss. It does not raise blood glucose or insulin levels, so it is recommended for diabetics. It is good for dental health because it does not promote tooth decay, as sugar does (although it does not prevent it either). For all these reasons, sucralose has surpassed aspartame as the most widely used artificial sweetener worldwide. It is currently used in over 6,500 products.

While sucralose appears to be preferable to aspartame, it should not be considered healthy or even benign. Some of the first complaints received after the sweetener hit the market were a variety of digestive troubles. A study by researchers at Duke University using lab animals found that sucralose altered normal gut bacteria, changed the pH levels in the intestines, increased body fat storage (promoting weight gain), and reduced the detoxification efficiency of the gastrointestinal tract.[11]

In another study at the University of North Florida, researchers discovered that sucralose inhibits the growth of Bacteroidetes bacteria in the digestive tract, altering the normal ratio of Bacteriodetes to Firmicutes.[12] Enriching the gastrointestinal tract with Firmicutes is associated with increased fat storage, weight gain, insulin resistance, inflammation, leaky gut, allergies, and digestive disorders, which would explain many of the symptoms reported by sucralose users, and why studies show that the use of sucralose promotes weight gain rather than weight loss.

Artificial sweeteners, especially sucralose, are believed by some researchers to be a major cause of inflammatory bowel disease (IBD), which includes ulcerative colitis and Crohn's disease. The incidence of IBD has risen dramatically with the introduction of saccharin and sucralose.[13] Twenty-five years ago IBD was most prevalent in the US, UK, and northern Europe. The incidence in Canada, however, was very low, only about half that of the United States. For example, the prevalence in Alberta, Canada in 1981 was only 44 per 100,000 population, compared with 91 per 100,000 in Olmsted County, Minnesota. By the year 2000 the prevalence of Crohn's disease in Canada had increased over six-fold; in Alberta it jumped to 283 per 100,000, while in Olmstead County it was only 174 per 100,000. By 2011 Canada had the highest incidence of IBD in the world. Why the dramatic jump in IBD in Canada? In 1991 Canada became the first country to approve the use of sucralose, leading some researchers to suspect that the dramatic rise in IBD is due to the earlier use of sucralose in Canada than in the US and Europe.[14]

Digestive disturbances are not the only concern with sucralose. One study linked large doses of sucralose to DNA damage in the intestines of mice.[15] Another study showed an increased risk of malignant tumors in rats at dosage levels proportionate to those consumed by humans.[16]

Sucralose has been shown in lab animals to cause bowel enlargement, kidney mineralization, abnormal pelvic tissue changes, migraines, and reduced weight of the spleen and thymus.[17-19] If it does all this to lab animals, it is likely to be having some undesirable effects on us as well. Indeed, people have reported a variety of adverse side effects with using sucralose, including abdominal pain, bloating, blurred vision, tremors, skin rashes, fatigue, dizziness, swelling, muscle cramps, joint pain, depression, migraines, and panic attacks, to name just a few. It is apparent that sucralose is not a safe sugar substitute.

NEOTAME

Neotame is a non-caloric sweetener made by the NutraSweet Company. It is one of the rare artificial sweeteners that was purposely developed as a sweetener rather than being discovered by accident. Chemically it is very similar to aspartame, though it was designed to overcome some of the problems associated with aspartame. Like aspartame, it is composed of aspartic acid, phenylalanine, and methanol. It has an additional dimethylbutyl molecule attached that binds the aspartic acid with the phenylalanine, reducing the amount of free phenylalanine released in the body, thus eliminating the need for a warning on labels for those with PKU.

Neotame is chemically more stable than aspartame and moderately heat-stable, allowing it to be used in foods exposed to moderate temperatures, although it is used mostly in soft drinks, gelatin desserts, yogurts, gum, protein bars, and protein shakes.

It is extremely potent, having a sweetness about 8,000 times that of sucrose and 40 times that of aspartame. Because of its exceeding sweetness, only a tiny amount is needed to sweeten foods, thus contributing a negligible amount of calories.

The FDA approved neotame for general use in 2002. It was soon thereafter approved in Europe, Australia, and New Zealand. Neotame entered the market much more discreetly than the other artificial sweeteners, without any fanfare or notice; even now many consumers are totally unaware of it. Which makes you wonder: if neotame is better and healthier than aspartame, why was it not introduced with more publicity and added to more products? At this time, relatively few foods are sweetened with neotame, less than 200.

The neotame website states that it's safe for use by people of all ages, including pregnant or breastfeeding women, teens, and children; and unlike aspartame can be used in cooking. While the website for neotame claims there are over 100 company sponsored scientific studies to support its safety, they are not readily available to the public. Because of its low profile, few independent studies have been done to verify its safety, so its effect on health is generally unknown. But since it is very similar to aspartame, it could possibly produce many of the same health problems. Perhaps the company knows more than it's letting on about the potential harm neotame might cause, and just doesn't want to generate the publicity that might lead to independent studies.

ADVANTAME

Advantame is a non-caloric sweetener produced by Ajinomoto, a Japanese company which is the world's largest maker of aspartame. Advantame is a derivative of aspartame and is similar in structure to neotame, with a sweetness that is up to 20,000 times that of sucrose, depending on how it is used. It was approved by the FDA in 2014 for general use in foods and beverages, except meat and poultry.

Like neotame, advantame's introduction into the marketplace has been very subtle, and currently few products contain it. Animal studies sponsored by Ajinomoto have not reported any evidence of carcinogenicity or toxicity.[20] However, since it is very similar to aspartame, there is concern that it may possess some of the same detrimental health effects. Also, any man-made substance that is thousands of times sweeter than sugar truly is more like a drug than a food, with possible unknown pharmaceutical effects.

Simply because the FDA approves a product doesn't make it safe. Based on company produced safety studies, the FDA has approved a multitude of food additives and drugs that have later been found to cause numerous health problems. For example, the pain relief medication Vioxx passed FDA screening for safety and approval, yet ended up taking 55,000 lives. The approval process is much more stringent for drugs than it is for food additives, yet dangerous drugs do pass the screening process. It is far more likely for harmful food additives to get approval than for drugs. We should be exceedingly cautious with very high-intensity sweeteners such as advantame and neotame.

STEVIA

Stevia leaves were used by the Guarani Indians in Paraguay for generations to sweeten yerba mate, a local bitter tasting tea-like beverage. The leaves were also chewed for their sweet taste. The discovery of stevia by western science is credited to Swiss botanist, Moisés Santiago Bertoni in the late 19th century. While exploring the eastern forests of Paraguay, he noticed a "strange plant" being used by his native guides. He published the first scientific description of the new species in 1905, which he named *Stevia rebaudiana Bertoni*.

In 1931 French chemists isolated the predominant compounds that give stevia its sweet taste. The product was a pure white crystalline powder they named stevioside. This new sweetener was discovered by enterprising Japanese businessmen in the 1960s, and in the 1970s was introduced into Japan by a consortium of food-product manufacturers. Japan is a world leader in the production and exportation of food additives that include stevia, aspartame, sucralose, acesulfame K, MSG, monoammonium glutamate (MAG), soy protein isolate, hydrolyzed vegetable protein, and others. Japan is the world's leading producer of aspartame, turning out 14,000 metric tons yearly, about 40 percent of all the aspartame sold worldwide. It is no wonder that they were the first to see the financial potential in stevia extract, and to begin marketing it

The fact that the Japanese have been using crystalline stevia powder since the 1970s is cited as proof of its safety. But long-time use of a product does not prove its safety. For example, Japan was also the first country to use and market monosodium glutamate (MSG), a flavor enhancer used in a wide variety of packaged, prepared foods. MSG is an excitotoxin, which means it can stimulate cell activity to the point of damage or death. Brain cells are particularly vulnerable. In fact, researchers will purposely administer MSG to lab animals to simulate the damage done by neurodegenerative diseases such as Alzheimer's. The first clues that MSG was harmful surfaced in the 1950s when researchers gave it to mice and it destroyed the retinas of their eyes and caused brain damage.[21] MSG has generated nearly as many consumer complaints as aspartame, with symptoms ranging from headaches and seizures to heart palpitations and chest pain. MSG is frequently used in Asian foods, giving rise to the term "Chinese Restaurant Syndrome" to describe the deleterious short-term symptoms from consuming food containing the additive.

MSG has been in use since 1909, far longer than stevia, and yet despite numerous studies documenting its destructive action and the multitude of consumer complaints, it is still used as a food additive worldwide. Simply because it has been used in Japan for over a century doesn't make it healthy.

By the 1980s a small amount of stevia leaf was being imported to the United States and Europe. Because of its herbal flavor, it was used almost exclusively to sweeten teas. In 1991 the FDA banned stevia leaf after studies found that it might be mutagenic and carcinogenic. Over the next several years a number of studies appeared that suggested it might also adversely affect reproductive health, liver and kidney function, and blood glucose metabolism. After doing their own investigation, the herb was banned in Europe eight years later and in numerous other countries worldwide.

The passage in 1994 of the Dietary Supplement Health and Education Act allowed stevia to be sold as an herbal dietary supplement, but not as a sweetener or food additive. In 2008 the FDA approved purified steviol glycosides for use as food additives; however, the ban on stevia leaf remained. Although it may seem contradictory, the FDA does not consider steviol glycosides to be the same as the stevia leaf. Steviol glycosides, they say, are "purified chemicals," not stevia. Following suit, Australia and New Zealand approved the sweetener in 2008, followed by the European Union in 2011 and Canada in 2012, in addition to many other countries.

More recently, additional concerns have arisen about steviol glycosides and all other low-calorie sweeteners. Studies show that regardless of the chemical makeup of a substance, if it has a sweet taste without the corresponding calories, it can stimulate sweet (sugar) addiction, encourage weight gain, promote insulin resistance, interfere with hormone regulation, and alter the gut microbiome.

Stevia has been heavily promoted as a harmless, even healthy, herbal sweetener with the sweetness of sugar but without any of the health risks. A natural sweetener that is harmless, as well as healthy is a dream come true for those looking for better alternatives to sugar and sugar substitutes. In fact, it seems too good to be true. As the saying goes, "If it sounds too good to be true, it probably is." This is apparently the case with stevia as well.

ACCEPTABLE DAILY INTAKE

Food additives are used in such small amounts that many people consider them insignificant, with few if any health consequences. Animal studies showing adverse effects to artificial sweeteners may have used dosages that were, proportionately, many times greater than what a human would ever consume. Although artificial sweeteners could cause problems at very high doses, we are told, at recommended levels of consumption they are perfectly safe. So what are these "recommended" levels of consumption? Most people aren't even aware that there are any recommended limits of consumption.

Government reviewing committees consisting of groups of scientists review all of the safety studies on specific food additives,

ADI for FDA Approved Sweeteners

Sweetener	ADI (mg/kg body weight)	Number of packets to equal ADI
Acesulfame K	15	23
Advantame	32.8	4,920
Aspartame	50	75
Neotame	0.3	23
Saccharin	15	45
Stevia	4	9
Sucralose	5	23

The ADI mg per kg body weight, is based on lean body mass—a person's proper or healthy weight for their height. You can view height and weight charts online at http://www.healthchecksystems. com/heightweightchart.htm. The number of tabletop sweetener packets needed to reach each ADI on this chart is based on a 60 kg (132 pound) person.

Source: http://www.fda.gov/Food/IngredientsPackagingLabeling/ FoodAdditivesIngredients/ucm397725.htm

such as artificial sweeteners, and make a determination as to the maximum amount of the substance that can be consumed daily without causing any appreciable harm. This amount is referred to as the acceptable daily intake (ADI). The ADIs for all of the approved artificial sweeteners have been determined and are listed in the table above, along with the equivalent amount in tabletop sweetener packets.

The larger the ADI number, the safer the sweetener is considered, and the greater is the supposed amount that can be safely consumed daily; conversely, the smaller the ADI, the greater is the supposed risk. Notice that, with the exception of neotame, stevia has the lowest ADI value of all the artificial sweeteners, meaning the scientific committees reviewing the studies have determined that stevia extract poses a greater health risk than do all the other artificial sweeteners. Keep in mind that these values weren't simply pulled out of a hat, but were determined after evaluating all available safety studies. Apparently, there are more concerns about the health effects of using stevia then there are for the other artificial sweeteners, including saccharin and aspartame.

Each packet of non-caloric sweetener is as sweet as 2 teaspoons of sugar. A 12-ounce (340 g) serving of sugar sweetened soda contains about 10 teaspoons of sugar. For the same level of sweetness, a sugar-free soda would have to contain the equivalent of 5 packets of artificial sweetener. From the table above we see that the daily limit for aspartame is 75 packets, which would be equivalent to fifteen 12-ounce cans of soda. The limit for stevia, on the other hand, is just 9 packets, or 2 cans of soda. How many people drink more than this daily?

Studies often show cancers or other adverse effects when animals consume large amounts of sweeteners, typically the equivalent of what is contained in 50 or more cans of diet soda a day. The ADI is set far below this with the assumption we would never consume that much soda, so there would be little risk. But sodas are not the only ways we are exposed to artificial sweeteners, they are used in a very large and growing number of other products, including coffee, tea, fruit juice, gum, salad dressings, candy, vitamin and herbal dietary supplements, cough syrup, pancake syrup, flavored waters, canned fruit, gelatin, ketchup; the list goes on and on. Consuming these other products daily, you could easily get the amount of sweetener equivalent to that in 50 cans of soda.

It is common to combine different non-caloric sweeteners to enhance the sweetness and minimize any aftertaste. This is true for stevia as well. Stevia is often combined with erythritol, sucralose, dextrose, and other sweeteners.

When you combine two or more drugs or other chemicals, their effect on the body can be very different from that which each has separately. That is why your doctor needs to know what medications you are taking before he can safely prescribe any additional ones. The same can be true with artificial sweeteners. Sweeteners do interact with each other, and since combining them intensifies their sweetness, it may also intensify their adverse reactions, or even cause an entirely new effect. We don't know what effects combining sweeteners may have because there have been no safety studies done testing multiple sweeteners. For all we know, a combination of two sweeteners, each with an ADI of 15 mg/kg, may become harmful when just 2 mg/kg is consumed. In other words, when you consume foods with multiple artificial sweeteners you are playing Russian roulette with your health. Any substance that has generated enough concern to be assigned an ADI value is better off left alone.

10

Sugar Alcohols and Monk Fruit

SUGAR ALCOHOLS

Sugar alcohols are a group of closely related chemicals with varying degrees of sweetness. They are found naturally in tiny amounts in some fruits, vegetables, and wood. Despite their name, they are neither sugars nor alcohols. They are carbohydrates with chemical structures that partially resemble both sugar and alcohol. Sugar alcohols are usually, but not always, identified by the suffix "-itol," as in xylitol, mannitol, and erythritol.

There are some notable differences between sugar alcohols and artificial sweeteners. Unlike the "nonnutritive" artificial sweeteners, sugar alcohols are, like sugars, considered "nutritive" sweeteners because they do provide calories when consumed. However, they contain fewer calories than sugar. Sugar provides 4 calories/gram, while sugar alcohols provide an average of 2 calories/gram (ranging from 1.5 to 3 calories/gram); therefore, they are more accurately described as low-calorie sweeteners. You may find some packaged products sweetened with sugar alcohols that claim zero calories. This isn't completely true. Remember, a product that provides less than 5 calories per serving can be labeled as zero-calorie.

Sugar alcohols are not "high-intensity" sweeteners like artificial sweeteners. The sweetness of sugar alcohols is much less than that of artificial sweeteners and generally even less than sugar.

134

The sweetness of sugar alcohols varies from 25 to 100 percent of that of sucrose. Depending on the type of sugar alcohol used, a greater volume may be required to equal the sweetness of sugar. But because sugar alcohols have fewer calories than sugar, their overall effect is of equal sweetness with fewer calories. They are often combined with other sweeteners to intensify the sweetness while reducing both calories and potential side effects.

It must be kept in mind that sugar alcohols are not calorie-free. If consumed in excess, their calories can add up and promote weight gain, raise blood sugar levels, and cause much of the same effects as sugar does. Sugar-free products that are sweetened with sugar alcohols may still contain a lot of carbohydrate and calories, and may therefore, be unsuitable for dieters and diabetics.

Although sugar alcohols are carbohydrates, they are only partially digested and absorbed as they travel through the digestive tract. This is why they provide fewer calories than sugar.

In the GI tract the presence of sugar alcohols draws water into the colon. This can alter gut pH and lead to the fermentation of bacteria and yeast. For this reason, high intakes of foods containing these sweeteners can lead to severe stomach cramping, bloating,

Common Sugar Alcohols

Sugar Alcohol	Calories/Gram	Sweetness Compared to Sucrose (%)
Erythritol	0.2	60 to 80
HSH*	3.0	25 to 50
Isomalt	2.0	45 to 65
Lactitol	2.0	30 to 40
Maltitol	2.1	90
Mannitol	1.6	50 to 70
Sorbitol	2.6	50 to 70
Xylitol	2.4	100

*Hydrogenated starch hydrolysates

gas, and diarrhea. Some sugar alcohols are more prone to digestive disturbances than others. Any food that contains sorbitol or mannitol must include a warning on its label stating: "Excess consumption may have a laxative effect." The American Diabetic Association advises that intakes greater than 50 grams/day of sorbitol or greater than 20 grams/day of mannitol may cause diarrhea. All of the sugar alcohols can have this effect. In the European Union they are banned from soft drinks because of their laxative effect. But each person is affected differently; some can consume more than the above threshold amounts without suffering problems, while others may consume far less and yet have severe reactions.

Regular use of these sweeteners reportedly allows the body to build a tolerance for them and somewhat reduce their side effects. A few studies indicate that xylitol can cause a shift in the gut microbiome in both animals and humans.[1] In human subjects, a measurable change in the gut microbiome has been observed after a single 30 gram oral dose of xylitol.[2] The decrease in symptoms after continued regular use is likely caused by an increase in the type of bacteria that are more efficient at digesting and extracting calories from the sugar alcohols, which likely increases the ratio of Firmicutes to Bacteroidetes.

Children are more vulnerable than adults to the effects of sugar alcohols, and should be more careful of the amounts consumed. Because sugar alcohols can cause water loss and digestive distress, they should not be consumed by anyone who suffers from digestive disorders such as irritable bowel syndrome (IBS), ulcerative colitis, Crohn's disease, celiac disease, and gluten intolerance, or who has the flu or other infections that cause dehydration or diarrhea, as they can make the condition worse.

Sugar alcohols can promote dehydration and the loss of electrolytes. Athletes and other physically active people should avoid sugar alcohols especially before or during exercise or competition, as these products will increase their risk of dehydration, muscle cramping, and heat stroke.

People on ketogenic diets for weight loss or other health reasons should avoid all sugar alcohols, as these sweeteners will make it much more difficult for them to get into nutritional ketosis. If they are already in ketosis, the sweeteners will quickly kick them out. Sugar alcohols are potent anti-ketogenic substances; you don't even need to swallow them for their anti-ketogenic effect to kick in. Simply chewing gum or using toothpaste or mouthwash containing sugar alcohols, is enough to prevent or reverse nutritional ketosis. As with stevia, the sweet taste alone is enough to activate hormones that produce the anti-ketogenic effect.

The most successful treatment for epilepsy is a ketogenic diet. This diet alone, without medication, can significantly reduce seizure frequency, and in many cases bring about a complete reversal of the condition. Sugar alcohols interfere with the healing effects of the ketogenic diet, and are known to increase seizure frequency.

Promoters of sugar alcohols claim that these sweeteners are safe and even health-promoting because they occur naturally in many plants. But we only get trace amounts in natural foods, generally not enough to have any significant effect. In processed foods we are exposed to quantities that are as much as 100 times greater than what we would normally get eating whole foods.

Our bodies are not designed to process such huge amounts of sugar alcohols, as evidenced by the stomach cramping and diarrhea they cause. That alone should tell you something is amiss. Flushing the GI tract with water to cause diarrhea is a process the body uses to expel unwanted substances. This process also affects digestive function, nutrient absorption, electrolyte balance, intestinal pH, and the gut microbiome.

Advertising for sugar alcohols often claims they are "all natural" or that they come from fruit, as if the sweeteners were simply extracted from fresh produce. While sugar alcohols occur naturally in some fruits and vegetables, these sources don't contain enough to be economically practical. Instead, they are commercially manufactured by a process involving some combination of hydrogenation, hydrolysis, and fermentation of sugar, corn, corncobs, and wood pulp, depending on the type of sugar alcohol being produced. The point is, sugar alcohols are not simple extracts, but manufactured products like most other food additives.

As with other sweeteners, a variety of health claims have been made about sugar alcohols. A number of studies have suggested that various sugar alcohols may help prevent dental cavities, increase collagen synthesis, thicken the skin, strengthen bones, prevent age-related skin damage and bone loss, prevent ear infections, and improve endothelial function in diabetics.[3-6] It's curious that a substance that supplies little to no calories, has no nutritional value, and is not well broken down or digested but passes out of body without change, could have such remarkable health-promoting effects on the body. A number of researchers have wondered the same thing and have questioned the validity of the health claims.[7]

The most thoroughly researched benefit of sugar alcohols, and specifically of xylitol, is their effect on dental health. Like other sugar substitutes, xylitol can replace sugar in the diet, reducing the risk of sugar-induced tooth decay. But some researchers believe it does more than just prevent cavities, that it actively fights the bacteria that cause them.

These researchers theorize that cariogenic bacteria (those that cause tooth decay) attempt to feed on xylitol by consuming it, but are unable to digest it and so expel it. This process drains energy from the organisms without supplying them with any energy. Repeated consumption and regurgitation of the xylitol causes the organisms to burn themselves out and essentially starve to death, thus reducing the amount of harmful bacteria in the mouth.[8]

The effectiveness of xylitol in combating tooth decay seems to vary with the dosage. The recommended amount of xylitol to prevent dental cavities is from 6 to 10 grams per day; any more or less than this decreases effectiveness. Preferably, xylitol should come from chewing gum or lozenges, so that it can bathe the teeth and allow the bacteria to go through the consumption and regurgitation cycle as often as possible. Doses of 3.4 grams/day or less have shown to be of no benefit in reducing oral bacteria.[9] A typical stick of gum contains between 0.3 and 0.4 grams of xylitol. That means you would have to chew 17 to 28 sticks of xylitol sweetened gum every day. That's a lot of gum chewing! Any less than this just won't cut it. Ten sticks of gum contain only 3.4 grams of xylitol, which provides little or no protection.

You can take larger doses in snack foods, but when you do the xylitol is quickly swallowed and provides no oral benefit.[10] Besides, any carbohydrate in the food also serves as food for the bacteria, preventing them from starving to death.

According to Philip Riley, PhD, and colleagues at the University of Manchester School of Dentistry in the UK, despite the number of published studies on xylitol's effect on dental health, there is little hard evidence that it actually fights tooth decay. Riley and his team analyzed the data on nearly 6,000 participants in 10 major studies and found the evidence to be very weak. He stated: "The evidence we identified did not allow us to make any robust conclusions about the effects of xylitol, and we were unable to prove any benefit in the natural sweetener for preventing tooth decay."[11] Although there have been a number of studies that suggest xylitol may help prevent tooth decay, it's only sure effect is to replace sugar that might otherwise promote cavities.

Sugar alcohols differ from one another principally by the number of carbons making up their carbon chain. Most have five or six carbons. Sorbitol has six carbons, xylitol has five, and erythritol has four. The simplest is the two carbon ethylene glycol, a sweet but notoriously toxic sugar alcohol used in antifreeze. Many a child and household pet have died after drinking sweet-tasting antifreeze. For this reason, manufacturers now make antifreeze with a bitter flavoring to prevent accidental poisoning. More complex sugar alcohols are generally nontoxic to humans, but not necessarily to animals or insects.

Among the sugar alcohol sweeteners, xylitol is the most notorious for its potentially toxic effects on animals. Many a veterinarian has had to save a dog's life or watch it die after it had consumed candy sweetened with xylitol. In some animals xylitol can have a dramatic effect by dropping blood sugar to dangerously low levels. Just a small amount of xylitol can be deadly to dogs, rabbits, ferrets, cows, goats, and baboons. Consuming candies or other human foods sweetened with xylitol can be enough to trigger the release of a huge amount of insulin into an animal's bloodstream, causing severe hypoglycemia that can lead to death. Independently of its effects on blood sugar, xylitol has also been reported to cause acute liver failure in dogs.[12]

Erythritol and to a lesser extent mannitol have been shown to kill fruit flies, and have been proposed as insecticides.[13] They don't die on contact as they do with some pesticides; the insects have to eat the sugar alcohols and die over a period of a few days.

We are told that erythritol is safe for human consumption. The physiology of a fruit fly is different from a human's, and what is dangerous to them may be perfectly safe for us. The same goes for xylitol and dogs. Maybe. But it just makes sense to be leery of any product that has demonstrated toxic effects on animals (including some primates) and that can also cause digestive distress in humans.

Erythritol is about 70 percent as sweet as sugar but provides only 0.2 calories per gram, making it a near zero-calorie sweetener. In the digestive tract, approximately 90 percent is absorbed into the bloodstream by the small intestine; however, most of this is eventually excreted unchanged in the urine. Because little is actually metabolized in the bloodstream, it does not have much affect on blood sugar. The 10 percent that is not absorbed passes through the colon and directly out of the body. Since little enters the colon, it normally does not have the laxative effect seen with other sugar alcohols, unless large amounts are consumed.

With all these potential benefits, erythritol is becoming one of the most popular sugar substitutes for dieters and diabetics. However, it may not be completely in the clear. The erythritol you buy in the store, and that is found in packaged foods, is not a natural sweetener that is simply extracted from fruit. It's a manufactured product derived from corn, and not ordinary corn but generally genetically engineered (GE) corn that is designed to withstand huge doses of pesticides. Unless it is labeled as "certified organic" it is almost certainly made from GE corn (at least in North America).

More troubling, however, is the fact that erythritol is much like an artificial sweetener, it has a sweet taste but no calories, and is not completely broken down by the body. As you have learned in previous chapters, these are the same characteristics that can fuel sweet addiction, increase hunger, sabotage weight loss efforts, promote insulin resistance, block ketone production, and alter gut microbiome. While erythritol appears to be a better choice than many other sugar substitutes, it may harbor some hidden dangers.

MONK FRUIT (LUO HAN GUO)

Monk fruit is a baseball size, sub-tropical melon that is cultivated in the mountains of southern China. According to legend, monk fruit was named after the Buddhist monks who first cultivated the fruit some 800 years ago. The Chinese name for monk fruit is luo han guo, sometimes spelled lo han kuo. "Luo han" means "monk" in Chinese, and "guo" means "fruit."

In some ways monk fruit sweetener is similar to stevia. It is derived from sweet-tasting, naturally occurring plant chemicals that are concentrated and purified. The botanical name of the fruit is *Siraitia grosvenorii*. The sweetness comes from a combination of fructose, glucose and mogrosides—a group of triterpene glycosides (saponins). The fruit contains five different mogrosides, numbered 1 to 5; the main sweetening component is mogroside 5, which is estimated to be 300 to 400 times sweeter than sucrose.[14]

Monk fruit is seldom used fresh because it has an unappetizing earthy or beany taste, tends to form off-flavors after harvest, and spoils quickly. The fruit contains sugars that cause rapid browning, fermentation, and flavor alteration.

Traditionally, the fruit was picked green, slowly dried in ovens until completely brown, and then stored dry until used. The drying process preserves the fruit and removes most of the objectionable flavors. However, drying also causes the formation of a bitter taste. For this reason, dried monk fruit and fruit juice extract have been restricted mostly for use in teas and soups in which sugar, honey, and other flavorings are added. Although sweet, the "off" flavors make the fruit useless as a general purpose sweetener.

In 1995 Procter & Gamble patented a process for removing the objectionable flavors. Most of these offending compounds contain sulfur, such as hydrogen disulfide, methionol, and dimethylsulfide, which come from sulfur containing amino acids. The sweet tasting mogrosides make up about 1 percent of the fresh fruit. Through solvent extraction and dehydration, a powder containing 80 percent mogrosides can be obtained with a sweetness that is about 200 times that of sucrose. Because it contains a small amount of sugar, each teaspoon serving supplies about 2 calories; sucrose has about 16 calories per teaspoon.

Based on a few animal studies, monk fruit appears to have little or no toxicity.[15] It received FDA GRAS status in 2009 and is approved as a food additive.

Studies suggest monk fruit may have anti-inflammatory, anti-diabetic, and anti-carcinogenic properties[16-19]

However, the sweetener is still relatively new and research has been limited. Few human studies have been done so its effect on human health is still pretty much unknown. In light of the fact that monk fruit is a high-intensity sweetener and nearly calorie-free, it seems likely that it would have the same drawbacks as stevia and other artificial sweeteners regarding sweet addiction, weight gain, insulin resistance, and alteration of the gut microbiome.

Currently, monk fruit is commercially grown only in the Guangxi province of southern China. Demand is high in both China and elsewhere, making it expensive and harder to find than other sweeteners. Presently, there are about 200 products that contain monk fruit, but this number is bound to grow as consumer demand and production capabilities increase in the near future.

11

Are Any Sweeteners Safe?

NO REASON TO USE
LOW-CALORIE SWEETENERS

Sugar has been identified as a major contributing factor in the explosion of obesity, diabetes, high blood pressure, and other health problems over the past two to three generations. Sugar substitutes have been promoted as a means to reduce sugar consumption while still allowing us to enjoy sweetened foods and beverages.

As I was uncovering the truth about stevia, I was reluctant to think of it as harmful. I believed that perhaps it might be of some benefit to some people in some situations. I knew it was definitely not for people following a ketogenic diet, because of its anti-ketogenic effect. But what about people on low-carb or low-calorie diets? Stevia has no calories and is low-carb, so could it be of help to them? The entire reason why people go on these diets is to lose excess weight, manage diabetes, or improve their health in some way. Stevia, as well as other low-calorie sweeteners, do just the opposite.

Like sugar, low-calorie sweeteners activate sweet receptors in our mouths and along our digestive tract; but unlike sugar, they provide little or no calories. This sets into motion a series of responses that affect how our bodies function to maintain energy balance. As a consequence, low-calorie sweeteners promote weight gain, increase insulin resistance (diabetes risk), and cause other health problems. All of the conditions why people use low-calorie sweeteners are made worse by their use. Therefore, there is no sensible reason to use stevia or any other sugar substitute.

Low-calorie sweeteners stimulate appetite, fuel sugar addiction, encourage fat storage and weight gain, promote insulin resistance and diabetes, and alter the gut microbiome (which can lead to inflammation, leaky gut, allergies, and digestive disorders). In addition, low-calorie sweeteners often have their own unique health issues ranging from migraines and dizziness to diarrhea and dermatitis. As a whole, they are no safer than sugar and in many ways far worse.

In addition to the low-calorie sweeteners described in this book, a few others are in development. New ones will likely appear in the near future promising to give you that sweet taste you crave without the calories or the health issues associated with previous sweeteners; some may even promise spectacular health benefits beyond blood sugar control and weight loss. Ignore them all. They are fighting a battle against nature and against basic physiology of the human body, a battle they can never win.

CHASING A DREAM

Over the years, low-calorie sweeteners have become enormously popular. An entire diet industry has been built up around low-calorie foods and meal plans that rely heavily on low-calorie sweeteners. The sweet taste is addicting and often overpowering. Giving up sugar and sweets is very difficult even if only for a brief period of time. Despite good intentions to lose weight and eat better, many people fail at this goal because of their addiction to sweets.

Most sugar substitutes provide little or no calories because they are not completely broken down or absorbed, but pass through the body mostly undigested, and presumably without any adverse effects. Since these substances simply go in and out of the body without change, they are believed to cause no harm. This offers the opportunity to eat sweets and desserts to our heart's content without any consequences—a virtual dream come true! You can have your cakes, cookies, fruited yogurts, snack bars, pancakes, and muffins without worry, because with low-calorie sweeteners it is possible for you to eat these types of foods without all the calories.

The first artificial sweetener, saccharin, opened the door to the idea that we could indulge in high-calorie foods without the calories. However, despite the fact that it was not absorbed well, when it was discovered that saccharin might cause cancer, people turned to cyclamate. But that proved to be worse than saccharin. Then came acesulfame K and aspartame. They were immediately embraced as better choices, but fell under suspicion as health issues began to arise. Then came sugar alcohols and sucralose, both of which were touted as healthier options to previous sweeteners, but their safety has become suspect too. When stevia appeared on the scene, it was immediately embraced as a wholesome, natural sweetener. The long awaited "safe" sweetener had finally arrived. But that image is now crumbling too. Indeed, every time a new sweetener comes along, it is promoted as safer than the last, but later is discovered to be not so wonderful after all. We keep chasing the dream of a safe and healthy no-calorie sweetener, but it continues to elude us.

If you are looking for a magic bullet that will allow you to eat sweetened foods with abandon, you are truly in a dream world. It isn't going to happen. With the discovery that any substance that has a sweet taste without the corresponding calories can cause more harm than sugar, it is unlikely that we will ever find a sweetener that can be consumed without any adverse consequences.

You need to have a change of mind. Once you realize that no sugar substitute will help you lose weight and improve your health, you can stop wasting your time chasing a dream that does not exist. You can focus on methods that do work, but require a new way of thinking about weight loss, and what constitutes a healthy diet.

JUNK FOOD BY ANY OTHER NAME IS STILL JUNK FOOD

We are told that non-caloric sweeteners are safe, and even if we assume that they are, that still doesn't mean they are healthy. Synthetic food dyes, MSG, chemical preservatives, even pesticide residue are considered safe as well, but they are not healthy. Each of these chemicals are known to cause serious health problems at some level; nevertheless, they are considered safe because only a small amount is added to any one food product. But we are often exposed to these substances through multiple products on a daily basis, often receiving far more than recommended limits. Long-term exposure to multiple sources of these so-called safe products is not safe.

Food manufacturers aren't satisfied with saying their sweeteners are safe, but go a step further and claim they are healthy because they help prevent tooth decay, control blood sugar levels, or reduce calorie intake. This, however, is true only because the sugar has been removed from your diet, not because the sweetener has been added. The sweeteners in themselves have no health-promoting properties. In fact, as you have seen in this book, they cause many of the same problems as sugar does and more, making them even worse than sugar.

The view of food manufacturers and their allies in government health organizations is captured in a statement from National Health Service (UK) dietitian Emma Carder: "Research into sweeteners shows they are perfectly safe to eat or drink on a daily basis as part of a healthy diet."[1]

Ahhhh...did you notice the glaring caveat in this statement? Artificial sweeteners are perfectly safe to consume "...as a part of a *healthy* diet" (my emphasis). What types of foods are artificial sweeteners used in? They are found almost exclusively in junk foods: soda, ice cream, candy, cakes, cookies, jellies, syrups, etc. None of these could remotely be called healthy. If you eat junk foods, that means you are displacing other, healthier foods like fresh fruits and vegetables from your diet to make room for nutritionally poor, additive-laden foods. In other words, if you eat foods sweetened with artificial sweeteners, you are *not* eating a healthy diet. Therefore, based on the statement from the National Health Service, we can assume artificial sweeteners are not safe to eat on a daily basis.

That's the problem with all low-calorie sweeteners, whether aspartame, erythritol, or stevia; they are generally used to sweeten junk foods and beverages that have no place in a healthy diet. Junk foods are those that are highly processed, contain few vitamins and minerals, and are often loaded with chemical additives. They can also be loaded with bad fat, refined carbohydrates, and sugar calories—empty calories with no nutritional benefit.

Removing the sugar and adding chemical sweeteners does not transform junk into gold. You don't automatically turn a candy bar (or protein bar) into a health food simply by replacing the sugar with a non-caloric sweetener. Junk foods by any other name are still junk foods. No amount of spin from the food manufacturers is going to change that.

There are many studies out there that try to make it appear that stevia, erythritol, or monk fruit have special health properties beyond their sweetening effect, in order to create the image they are healthy. But the fact remains these sweeteners are always used in sweets, desserts, and junk foods. You will never be successful losing excess weight, controlling diabetes, or improving your health eating these types of foods. You will only find success by changing your diet and the types of foods you eat.

THE BEST KEPT WEIGHT LOSS SECRET

According to the Mayo Clinic, 95 percent of those people who go on low-calorie diets to lose weight, regain all their weight within five years. That's a 95 percent failure rate! Only 5 percent manage to keep the weight off long-term. Why such a dismal success rate? The number one reason why people fail to lose weight and keep it off is because of sweet addiction. When they go on a diet, they continue to feed their sweet tooth. It's not all their fault though, they are encouraged to do it.

When you go on a diet, you are told that you can eat pancakes with sugar-free syrup, muffins, energy bars, low-fat ice cream, sugar-free soda and other "diet" foods sweetened with sugar substitutes. But what invariably happens? Losing weight is a real struggle. You suffer with starvation and constant hunger. Some weight may come off initially, but before long it all comes right back. You followed the diet to the letter but you just could not keep it off. What went wrong?

Regardless of the type of diet chosen—low-fat, low-carb, high-protein, ketogenic, etc.—dieters are encouraged to eat foods sweetened with sugar substitutes to help them in their weight loss efforts, but it is precisely these types of foods that prevent them from achieving lasting success.

If you continue to eat the same types of foods that made you gain weight in the first place, regardless of whether they are sweetened with sugar or with a substitute, they are going to make you fat and prevent you from achieving your weight loss goals. Replacing the sugar with low-calorie sweeteners won't change anything. Successful dieting is not a simple adjustment in one ingredient, but requires a change in your way of thinking about foods and the types of foods you eat.

You will never be successful losing weight, controlling your blood sugar, or improving your health by using low-calorie sweeteners. Despite the claims, none of them have ever proven to be effective in helping people lose weight.

People don't get fat by eating zucchini, cauliflower, or mushrooms. They don't get fat by eating steak, eggs, or even cheese. It is very difficult to get fat eating natural, nutrient-rich foods like these. What makes people fat is eating candy, pies, cakes, ice cream, cookies, crackers, and chips—junk foods! This is especially true when these foods have been sweetened with sugar substitutes.

If you want to find success losing excess weight, managing your blood sugar, and improving your overall health, choose a diet built around whole, natural foods. This is the secret to successful weight loss. A diet that relies on the natural sweetness of fresh foods, without sugar and without sugar substitutes that will free you from the chains of sweet addiction, yet will still allow you to enjoy satisfying, delicious meals. It is possible.

If you have been using stevia or any other nonnutritive sweetener for any period of time, your GI tract could be in serious trouble. You may have tried all types of weight loss diets with limited or no success. The reason isn't that you didn't follow the program properly, or that you didn't have the willpower to stick it out. The problem could be that your GI tract is overpopulated with the wrong types of bacteria, having more Firmicutes than Bacteroidetes. The more Firmicutes you have, the fatter and sicker you get. The solution to this problem is simple—a sugar-free, and sugar substitute-free, natural foods diet.

A natural foods diet is one that focuses on eating whole, natural foods such as fresh fruits in season, fresh and fermented vegetables, unprocessed meats, eggs, full-fat dairy, nuts, good fats, and even some whole grains if appropriate, preferably all organically grown or raised. It could be described as a traditional diet such as that promoted by the Weston A. Price and Price-Pottenger Foundations; it could be a Paleo diet, or a low-carb or even a ketogenic diet. There are many options.

I have found that a vegetable-based, low-carb, high-fat ketogenic diet to be ideal for reestablishing the gut microbiome and shedding excess weight. The diet has also proven very helpful in controlling diabetes and treating a number of health issues such as epilepsy, Alzheimer's disease, Parkinson's disease, heart disease, digestive disorders (Crohn's, ulcerative colitis, etc.) and others. For more information about this diet see my book *The Coconut Ketogenic Diet: Supercharge Your Metabolism, Revitalize Thyroid Function, and Lose Excess Weight*. Need a source of tasty recipes, without artificial sweeteners? See my book *Dr. Fife's Keto Cookery: Nutritious and Delicious Ketogenic Recipes for Healthy Living*.

THE BEST SWEETENERS

There is no such thing as a "healthy" sweetener, and probably never will be. Some sweeteners are better than others, but none could be termed healthy. Believe it or not, sugar, or at least some forms of it, is a better option than any of the low-calorie sugar substitutes.

Our bodies were programmed to associate the sweet taste with a corresponding amount of calories. Our bodies know how to process and utilize sugar to produce energy. In fact, every cell in your body knows how to metabolize sugar; it's the primary fuel that keeps us alive. Sugar is found in all plants. It is an essential component of mother's milk. It is natural part of the human diet. That is more than you can say for any of the sugar substitutes. The primary problem with sugar is that we have discovered how to refine it and concentrate it and use it to make treats and desserts. Sugary foods are around us everywhere. It is far too easy to become addicted to sugar, and many of us are. It is the addiction, and overconsumption, that makes sugar so detrimental.

On average we each consume about 150 pounds (68 kg) of sugar a year. That amounts to over 44 teaspoons of sugar a day! This is added sugar, the sugar we add to coffee and cereal, as well as what the sugar manufacturers add to foods and beverages. It does not include the sugars that occur naturally in foods such as fruits and vegetables. No wonder sugar is associated with so many health problems. Anything consumed in huge amounts can be harmful.

If we ate as much low-calorie sweeteners as we do sugar, the consequences would be far worse. Two teaspoons of sugar equals the amount of sweetness in one packet of low-calorie sweetener. If you replaced all the added sugar in the typical diet (44 teaspoons) with a low-calorie sweetener, you would consume the equivalent of 22 packets of sweetener. That's 22 packets of stevia or sucralose or saccharin. From the table on page 131 we see that the so-called safe limit for stevia is 9 packets, for sucralose 23, and for saccharin 45. We know that these sweeteners can cause health problems at levels far below these limits; and when sweeteners are combined we have no idea what effect they have on the body, as no studies have been done to evaluate the synergistic effect of multiple sweeteners. Our bodies, at least, can metabolize sugar to produce energy, we can't say the same about chemical sweeteners. They serve no useful purpose in the body, and tend just to cause trouble.

The US Department of Agriculture (USDA) recommends that we limit our added sugar intake to no more than 6 percent of total calories consumed. For a typical 2,000 calorie diet, that would equate to 32 g or 8 teaspoons of sugar per day.[2] The American Heart Association recommends less than 10 percent of total calories.[3] That would be far below the average, and a sensible limit for most people. If you have diabetes it would be wise to eliminate all sugar from your diet, including the sugar that is added to processed foods, so you would need to read the Nutrition Facts panel on the package to know how much sugar there is in each serving. Unprocessed fresh foods don't have any added sugar. By eliminating low-calorie sweeteners and reducing the amount of sugar you eat, you will also limit your consumption of processed poor-quality foods, which will help tremendously in your quest to lose weight and gain better health.

The type of sugar you eat is also important. Natural or minimally processed sugars are preferred over highly processed ones. Fructose is one of the most highly processed, and is absolutely the worst in terms of what it does to your health. Fructose promotes obesity, liver disease, diabetes, heart disease, and other health problems more than any other sugar. Fructose is just as bad for you as any of the non-caloric sweeteners. Avoid all high fructose sweeteners, including corn syrup, high-fructose corn syrup, and agave.

Natural sweeteners are somewhat better than refined sugar because they contain a small amount of vitamins and minerals, so they are not completely empty calories. The best natural sweeteners include maple sugar/syrup, coconut sugar/syrup, raw honey, dehydrated sugarcane juice (Sucanat, rapadura, panela, jaggery, and muscovado), date sugar, brown rice syrup, and barley malt syrup. If you must use something to sweeten your food, your best choice would be a small amount of one of these natural sweeteners.

References

Chapter 1: The Bittersweet Truth About Stevia

1. Fife, B. Stop *Alzheimer's Now!: How to Prevent and Reverse Dementia, Parkinson's, ALS, Multiple Sclerosis, and Other Neurodegenerative Disorders, Second Edition*. Piccadilly Books, Ltd.: Colorado Springs, CO; 2016.

2. Fife, B. *Stop Autism Now!: A Parent's Guide to Preventing and Reversing Autism Spectrum Disorders*. Piccadilly Books, Ltd.: Colorado Springs, CO; 2012.

3. Fife, B. *Stop Vision Loss Now!: Prevent and Heal Cataracts, Glaucoma, Macular Degeneration, and Other Common Eye Disorders*. Piccadilly Books, Ltd.: Colorado Springs, CO; 2015.

4. Fife, B. *The Coconut Ketogenic Diet: Supercharge Your Metabolism, Revitalize Thyroid Function, and Lose Excess Weight*. Piccadilly Books, Ltd.: Colorado Springs, CO; 2014.

5. Volek, JS and Phinney, SD. *The Art and Science of Low Carbohydrate Performance*. Beyond Obesity, LLC; 2012.

6. Daniel, KT. *The Whole Soy Story: The Dark Side of America's Favorite Health Food*. New Trends Publishing, Inc: Washington, DC; 2005.

7. Wölwer-Rieck, U. The leaves of Stevia rebaudiana (Bertoni), their constituents and the analyses thereof: a review. *J Agric Food Chem* 2012;60:886-895.

8. Matsui, M, et al. Evaluation of the genotoxicity of stevioside and steviol using six in vitro and one in vivo mutagenicity assays. *Mutagenesis* 1996;11:573-579.

9. Wasuntarawat, C, et al. Developmental toxicity of steviol, a metabolite of stevioside, in the hamster. *Drug Chem Toxicol* 1998;21:207-222.

10. Huxtable, R J. Pharmacology and toxicology of stevioside, rebaudioside A, and steviol in: *Stevia: The Genus Stevia*, Edited by A. Douglas Kinghorn. Taylor & Francis: London; 2002.

Chapter 2: The Problems with Low-Calorie Sweeteners

1. Olivier, B, et al. Review of the nutritional benefits and risks related to intense sweeteners. *Arch Public Health* 2015;73:41.

2. Ford, ES and Dietz, WH. Trends in energy intake among adults in the United States: findings from NHANES. *Am J Clin Nutr* 2013;97:848-853.

3. www.foodfacts.com. Accessed 7/8/2016.

4. Fowler, SP, et al. Fueling the obesity epidemic? Artificially sweetened beverage use and long-term weight gain. *Obesity* (Silver Spring, MD) 2008;16:1894-1900.

5. Forshee, RA and Storey, ML. Total beverage consumption and beverage choices among children and adolescents. *Int J Food Sci Nutr* 2003;54:297–307.

6. Hampton, T. Sugar substitutes linked to weight gain. *JAMA* 2008;299:2137–2138.

7. Swithers, SE., et al. General and persistent effects of high-intensity sweeteners on body weight gain and caloric compensation in rats. *Behav Neurosci* 2009;123:772-780.

8. Swithers, SE., et al. High-intensity sweeteners and energy balance. *Physiol Behav* 2010;100:55-62.

9. Stellman, SD and Garfinkel, L. Artificial sweetener use and one-year weight change among women. *Prev Med* 1986;15:195–202.

10. Blum, JW., et al. Beverage consumption patterns in elementary school aged children across a two-year period. *J Am Coll Nutr* 2005;24:93–98.

11. Mattes, RD and Popkin, BM. Nonnutritive sweetener consumption in humans: effects on appetite and food intake and their putative mechanisms. *Am J Clin Nutr* 2009;89:1-14.

12. Bruyere, O, et al. Review of the nutritional benefits and risks related to intense sweeteners. *Archives of Public Health* 2015;73:41.

13. De la Hunty, A, et al. A review of the effectiveness of aspartame in helping with weight control. *Nutrition Bulletin* 2006;31:115-128.

14. Miller, PE and Perez, V. Low-calorie sweeteners and body weight and composition: a meta-analysis of randomized controlled trials and prospective cohort studies. *Am J Clin Nutr* 2014;100:765-777.

15. Tordoff, MG and Friedman, MI. Drinking saccharin increases food intake and preference—I. Comparison with other drinks. *Appetite* 1989;12:1-10.

16. No author listed. Saccharin consumption increases food consumption in rats. *Nutr Rev* 1990;48:163-165.

17. Tordoff, MG. How do non-nutritive sweeteners increase food intake? *Appetite* 1988;11 Suppl1:5-11.

18. Lavin, JH, et al. The effect of sucrose- and aspartame-sweetened drinks on energy intake, hunger and food choice of female, moderately restrained eaters. *Int J Obes Relat Metab Disord* 1997;21:37–42.

19. King, NA, et al. Effects of sweetness and energy in drinks on food intake following exercise. *Physiol Behav* 1999;66:375–379.

20. Yang, Q. Gain weight by "going diet?" Artificial sweeteners and the neurobiology of sugar cravings. *Yale J Biol Med* 2010;83:101-108.

21. Blundell, J and Hill, A. Paradoxical effects of an intense sweetener (aspartame) on appetite. *Lancet* 1986;327:1092-1093.

22. Rogers, PJ, et al. Uncoupling sweet taste and calories: comparison of the effects of glucose and three intense sweeteners on hunger and food intake. *Physiol Behave* 1988;43:547-552.

23. Tordoff, MG and Alleva, AM. Oral stimulation with aspartame increases hunger. *Physiol Behav* 1990;47:555-559.

24. Blundell, JE and Hill, AJ. Paradoxical effects of an intense sweetener (aspartame) on appetite. *Lancet* 1986;1:1092-1093.

25. Rogers, PJ and Blundell, JE. Separating the actions of sweetness and calories: effects of saccharin and carbohydrates on hunger and food intake in human subjects. *Physiol Behav* 1989;45:1093-1099.

26. Tordoff, MG and Alleva, AM. Oral stimulation with aspartame increases hunger. *Physiol Behav* 1990;47:555-559.

27. Rogers, PJ and Blundell, JE. Separating the actions of sweetness and calories: Effects of saccharin and carbohydrates on hunger and food intake in human subjects. *Physiol Behav* 1989;45:1093-1099.

28. Black, RM, et al. Consuming aspartame with and without taste: differential effects on appetite and food intake of young adult males. *Physiol Behav* 1993;53:459–466.

29. Swithers, SE and Davidson, TL. A role for sweet taste: calorie predictive relations in energy regulation by rats. *Behav Neurosci* 2008;122(1):161–173.

30. Fildes, A, et al. Probability of an obese person attaining normal body weight: cohort study using electronic health records. *American Journal of Public Health* 2015;e1DOI:10.2105/AJPH.2015.302773.

31. Magalle, L, et al. Intense sweetness surpasses cocaine reward. *PLoS One* 2007;8e698.

32. Carroll, ME, et al. A concurrently available nondrug reinforcer prevents the acquisition or decreases the maintenance of cocaine-reinforced behavior. *Psychopharmacology*1989;97:23–29.

33. Carroll, ME and Lac, ST. Autoshaping i.v. cocaine self-administration in rats: effects of nondrug alternative reinforcers on acquisition. *Psychopharmacology* 1993;110:5–12.

34. Sciafani, A, et al. Stevia and saccharin preferences in rats and mice. *Chem Senses* 2010;35:433-443.

35. Hajnal, A, et al. Oral sucrose stimulation increases accumbens dopamine in the rat. *Am J Physiol Regul Integr Comp Physiol* 2004;286:R31–R37.

36. Mark, GP, et al. A conditioned stimulus decreases extracellular dopamine in the nucleus accumbens after the development of a learned taste aversion. *Brain Res* 1991;551:308–310.

37. d'Anci, KE, et al. Duration of sucrose availability differentially alters morphine-induced analgesia in rats. *Pharmacol Biochem Behav* 1996;54:693–697.

38. Colantuoni, C, et al. Evidence that intermittent, excessive sugar intake causes endogenous opioid dependence. *Obes Res* 2004;10:478–488.

39. Wang, GJ, et al. Similarity between obesity and drug addiction as assessed by neurofunctional imaging: a concept review. *J Addict Dis* 2004;23:39–53.

40. Wang, GJ, et al. Gastric stimulation in obese subjects activates the hippocampus and other regions involved in brain reward circuitry. *Proc Natl Acad Sci USA* 2006;103:15641–15465.

41. Fife, B. *The Coconut Ketogenic Diet: Supercharge Your Metabolism, Revitalize Thyroid Function, and Lose Excess Weight.* Piccadilly Books, Ltd.: Colorado Springs, CO; 2014.

42. Fife, B. *Stop Vision Loss Now!: Prevent and Heal Cataracts, Glaucoma, Macular Degeneration, and Other Common Eye Disorders.* Piccadilly Books, Ltd.: Colorado Springs, CO; 2015.

43. Hershline, R. *The New Face of Alcoholism Treatment Brain Energy Management.* Roger Hershline; 2015.

44. Volek, JS and Phinney, SD. *The Art and Science of Low Carbohydrate Performance.* Beyond Obesity, LLC; 2012.

45. Hubler, MO, et al. Influence of stevioside on hepatic glycogen levels in fasted rats. *Res Commun Chem Pathol Pharmacol* 1994;84:111-118.

Chapter 3: Health Claims

1. Satishkumar, J, et al. In-vitro antimicrobial and antitumor activities of Stevia rebaudiana (Asteraceae) leaf extracts. *Trop J Pharm Res* 2008;7:1143–1149.

2. Mohan, K and Robert J. Hepatoprotective effects of Stevia rebaudiana Bertoni leaf extract in CCl_4-induced liver injury in albino rats. *Med Arom Plant Sci Biotechnol* 2009;3:59–61.

3. Takahashi, K, et al. Analysis of anti-rotavirus activity of extract from Stevia rebaudiana. *Antiviral Res* 2001;49:15–24.

4. Gopalakrishnan, B, et al. Free radical scavenging activity of flavonoid containing leaf extracts of Stevia rebaudiana Bert. *Anc Sci Life* 2006;25:44-48.

5. Shivanna, N, et al. Antioxidant, anti-diabetic and renal protective properties of Stevia rebaudiana. *J Diabetes Complications* 2013;27:103-113.

6. Hsieh, MH, et al. Efficacy and tolerability of oral stevioside in patients with mild essential hypertension: a two-year, randomized, placebo-controlled study. *Clin Ther* 2003;25(11):2797-808.

7. Ferri, LA, et al. Investigation of the antihypertensive effect of oral crude stevioside in patients with mild essential hypertension. *Phytother Res* 2006;20:732-736.

8. Curi, R, et al. Effect of Stevia rebaudiana on glucose tolerance in normal adult human. *Braz J Med Biol Res* 1986;19:771-774.

9. Gregersen, S, et al. Antihyperglycemic effects of stevioside in type 2 diabetic subjects. *Metabolism* 2004;53:73-76.

10. Just, T, et al. Cephalic phase insulin release in healthy humans after taste stimulation? *Appetite* 2008;51:622-627.

11. Chang, JC, et al. Increase of insulin sensitivity by stevioside in fructose-rich chow-fed rats. *Horm Metab Res* 2005;37:610-616.

12. Chen, TH, et al. Mechanism of the hypoglycemic effect of stevioside, a glycoside of Stevia rebaudiana. *Planta Med* 2005;71:108–113.

13. Maki, KC, et al. The hemodynamic effects of rebaudioside A in healthy adults with normal and low-normal blood pressure. *Food Chem Toxicol* 2008;26:S40-S46.

14. Maki, KC, et al. Chronic consumption of rebaudioside A, a steviol glycoside, in men and women with type 2 diabetes mellitus. *Food Chem Toxicol* 2008;l46:S47-S53.

15. Barriocanal, LA, et al. Apparent lack of pharmacological effect of steviol glycosides used as sweeteners in humans. A pilot study of repeated exposures in some normotensive and hypotensive individuals and in Type 1 and Type 2 diabetics. *Requl Toxicol Pharmacol* 2008;51:37-41.

Chapter 4: Safety Concerns

1. Kinghorn, AD. Overview. In: Kinghorn, A.D. (Ed.), *Stevia: the Genus Stevia (Medicinal and Aromatic Plants - Industrial Profiles)*. Taylor & Francis/CRC Press: NewYork/London,UK;2002:1-17.

2. Mazzei-Planas, G and Kuc, J. Contraceptive properties of Stevia rebaudiana. *Science* 1968;162:1007.

3. Oliveira-Filho, RM, et al. Chronic administration of aqueous extract of Stevia rebaudiana (Bert.) Bertoni in rats: endocrine effects. *Gen Phamacol* 1989;20:187-191.

4. Melis, MS. Effects of chronic administration of Stevia rebaudiana on fertility in rats. *J Ethnopharmacol* 1999;67:157-161.

5. Halldorsson, TI, et al. Intake of artificially sweetened soft drinks and risk of preterm delivery: a prospective cohort study in 59,334 Danish pregnant women. *Am J Clin Nutr* 2010;92:626-633.

6. Maki, KC, et al. Chronic consumption of rebaudioside A, a steviol glycoside, in men and women with type 2 diabetes mellitus. *Food Chem Toxicol* 2008;46:S47-S53.

7. Curry, LL and Roberts, A. Subchronic toxicity of rebaudioside A. *Food Chem Toxicol* 2008;46 Suppl7:S11-S20.

8. Flamm, WG, et al. Long-term food consumption and body weight changes in neotame safety studies are consistent with the allometric relationship observed for other sweeteners and during dietary restrictions. *Regul Toxicol Pharmacol* 2003;38:144-156.

9. Pezzuto, JM, et al. Metabolically activated steviol, the aglycone of stevioside, is mutagenic. *Proc Natl Acad Sci USA* 1985;82:2478-2482.

10. Terai, T, et al. Mutagenicity of steviol and its oxidative derivatives in Salmonella typhimurium TM677. *Chem Pharm Bull* 2002;50:1007-1010.

11. Matsui, M, et al. Evaluation of the genotoxicity of stevioside and steviol using six in vitro and one in vivo mutagenicity assays. *Mutagenesis* 1996;11:573-579.

12. Matsui, M, et al. 1989. Detection of deletion mutations in pSV2-gpt plasmids induced by metabolically activated steviol. Selected abstracts of the 17th Annual Meeting of the Environmental Mutagen Society of Japan. Mutat Res 1989;216:353-385.

13. Suttajit, M, et al. 1993. Mutagenicity and human chromosomal effect of stevioside, a sweetener from stevia rebaudiana bertoni. Environ Health Perspect 1993;101Suppl:53-56.

14. Nunes, APM, et al. Analysis of genotoxic potentiality of stevioside by comet assay. *Food Chem Toxicol* 2007;45:662-666.

15. Hutapea, AM, et al. Digestion of stevioside, a natural sweetener, by various digestive enzymes. *J Clin Biochem Nutr* 1997;23:177-186.

16. Kobylewski, S. and Eckhert, CD. *Toxicology of Rebaudioside A: A Review.* UCLA Department of Environmental Health Sciences and Molecular Toxicology 2008:1-28.

17. Williams, LD and Burdock, GA. Genotoxicity studies on a high-purity rebaudioside A preparation. *Food Chem Toxicol* 2009;47:1831-1836.

18. Glinsukon, T, et al. Stevioside: a natural sweetener from Stevia rebaudiana Bertoni: toxicological evaluation. *Thai J Toxicol* 1988;4:1-22.

19. Akihisa, T, et al. Microbial transformation of isosteviol and inhibitory effects on Epstein-Barr virus activation of the transformation products. *J Nat Prod* 2004;67:407-410.

20. Yasukawa, K, et al. Inhibitory effect of stevoiside on tumor promotion by 12-O-tetradecanoylphorbol-13-acetate in two-stage carcinogenesis in mouse skin. *Biol Pharm Bull* 2002;25:1488-1490.

21. Compradre, CM, et al. Mass spectral analysis of some derivatives and in vitro metabolism of steviol, the aglycone of the natural sweeteners, stevioside, rebaudioside A, and rubusoside. *Biomed Environ Mass Spectrom* 1988;15:211-222.

22. Denina, I, et al. The influence of stevia glycosides on the growth of Lactobacillus reuteri strains. *Lett Appl Microbiol* 2014;58:278-284.

23. Terai, T, et al. Mutagenicity of steviol and its oxidative derivative in Salmonella typhimuriium TM677. *Chem Pharm Bull* 2002;50:1007-1010.

24. Melis, MS. Chronic administration of aqueous extract of Stevia rebaudiana in rats: renal effects. *J Ethnopharmacol* 1995;47:129-134.

25. Melis, MS. A crude extract of Stevia rebaudiana increases the renal plasma flow of normal and hypertensive rats. *Braz J Med Biol Res* 1996;29:669-675.

26. Toskulkao, C, et al. Acute toxicity of stevioside, a natural sweetener, and its metabolite, steviol, in several animal species. *Drug Chem Toxicol* 1997;20:31-44.

27. Atteh, JO, et al. Evaluation of supplementary stevia (Stevia rebaudiana, bertoni) leaves and stevioside in broiler diets: effects on feed intake, nutrient metabolism, blood parameters and growth performance. *J Anim Physiol Anim Nutr (Berl)* 2008;92:640-649.

28. Yang, Q. Gain weight by "going diet?" Artificial sweeteners and the neurobiology of sugar cravings. *Yale J Biol Med* 2010;83:101-108.

29. Wasuntarawat, C, et al. Developmental toxicity of steviol, a metabolite of stevioside, in the hamster. *Drug Chem Toxicol* 1998;21:207-222.

Chapter 5: Conflicting and Confusing Studies

1. Matsui, M, et al. Regionally-targeted mutagenesis by metabolically-activated steviol: DNA sequence analysis of steviol-induced mutants of guanine phosphoribosyltransferase (gpt) gene of Salmonella typhimurium TM677. *Mutagenesis* 1996;11:565-572.
2. White, J and Bero, L. Corporate manipulation of research: strategies similar across five industries. *Stanford Law Pol Rev* 2010;21:105-134.

Chapter 6: Digestive Health and Function

1. Abou-Donia, MB, et al. Splenda alters gut microflora and increases intestinal p-glycoprotein and cytochrome p-450 in male rats. *J Toxicol Environ Health A* 2008;71:1415-1429.
2. Suez, J, et al. Artificial sweeteners induce glucose intolerance by altering the gut microbiota. *Nature* 2014;514:181-186.
3. Denina, I., et al. The influence of stevia glycosides on the growth of Lactobacillus reuteri strains. *Lett Appl Microbiol* 2014;58:278-284.
4. Fagherazzi, G, et al. Consumption of artificially and sugar-sweetened beverages and incident type 2 diabetes in the Etude Epidemiologique aupres des femmes de la Mutuelle Generale de l'Education nationale-European Prospective Investigation into Cancer and Nutrition cohort. *Am J Clin Nutr* 2013;97:517-523.
5. Sakurai, M, et al. Sugar-sweetened beverage and diet soda consumption and the 7-year risk for type 2 diabetes in middle-aged Japanese men. *Eur J Nutr* 2014;53:251-258.
6. Qin, J, et al. A metagenome-wide association study of gut microbiota in type 2 diabetes. *Nature* 2012;490:55-60.
7. Shell, ER. Artificial sweeteners may change our gut bacteria in dangerous ways. *Scientific American* 2015;312(4).
8. Abbott, A. Sugar substitutes linked to obesity. *Nature* 2014;513(7518):290.
9. Ley, RE, et al. Microbial ecology: Human gut microbes associated with obesity. *Nature* 2006;444:1022-1023.
10. Ridaura, VK, et al. Cultured gut microbiota from twins discordant for obesity modulate adiposity and metabolic phenotypes in mice. *Science* 2013;341(6150):10.1126/science.1241214.

11. Backhed, F, et al. The gut microbiota as an environmental factor that regulates fat storage. *Proc Natl Acad Sci USA* 2004;101:15718-15723.

12. Turnbaugh, PJ, et al. Diet-Induced obesity is linked to marked but reversible alterations in the mouse distal gut microbiome. *Cell Host Microbe* 2008;3:213-223.

13. Turnbaugh, PJ, et al. An obesity-associated gut microbiome with increased capacity for energy harvest. *Nature* 2006;444:1027-1031.

14. Transande, L, et al. Infant antibiotic exposures and early-life body mass. *Int J Obes* 2013;37:16-23.

15. Blustein, J, et al. Association of caesarean delivery with child adiposity from age 6 weeks to 15 years. *In J Obes (Lond)* 2013;37:900-906.

16. Ogden,CL, et al. Prevalence of childhood and adult obesity in the United States, 2011-2012. *JAMA* 2014;311:806-814.

17. Shields, M. Overweight and obesity among children and youth. *Health Rep* 2006;17:27-42.

18. Azad, MB, et al. Association between artificially sweetened beverage consumption during pregnancy and infant body mass index. *JAMA Pediatr* 2016;170:662-670.

19. Araujo, JR, et al. Exposure to non-nutritive sweeteners during pregnancy and lactation: impact in programming of metabolic diseases in the progeny later in life. *Reprod Toxicol* 2014;49:196-201.

20. Englund-Ogge, L, et al. Association between intake of artificially sweetened and sugar sweetened beverages and preterm delivery: a large prospective cohort study. *Am J Clin Nutr* 2012;96:552-559.

21. Maslova, E, et al. Consumption of artificially-sweetened soft drinks in pregnancy and risk of child asthma and allergic rhinitis. *PLoS One* 2013;8:e57261.

22. Petersen, SB, et al. Maternal dietary patterns during pregnancy in relation to offspring forearm fractures: prospective study from the Danish National Birth Cohort. *Nutrients* 2015;7:2382-2400.

23. Nettleton, JA, et al. Diet soda intake and risk of incident metabolic syndrome and type 2 diabetes in the Multi-Ethnic Study of Atherosclerosis (MESA). *Diabetes Care* 2009;32:688-694.

24. Lutsey, PL, et al. Dietary intake and the development of the metabolic syndrome: The Atherosclerosis Risk in Communities Study. *Circulation* 2008;117:754-761.

25. Fujita, Y, et al. Incretin release from gut is acutely enhanced by sugar but not by sweeteners in vivo. *Am J Physiol Endocrinol Metab* 2008;296:E473-E479.

26. Perry, B and Wang, Y. Appetite regulation and weight control: the role of gut hormones. *Nutr Diabetes* 2012;16:e26.

27. Verdich, C, et al. The role of postprandial releases of insulin and incretin hormones in meal-induced satiety—effect of obesity and weight reduction. *Int J Obes Relat Metab Disord* 2001;25:1206-1214.

28. Jang, HJ, et al. Gut-expressed gustducin and taste receptors regulate secretion of glucagon-like peptide-1. *Proc Natl Acad Sci USA* 2007;104:15069-15074.

29. Rozengurt, E. Taste receptors in the gastrointestinal tract. I. Bitter taste receptors and alpha-gustducin in the mammalian gut. *Am J Physiol Gastrointest Liver Physiol* 2006;291:G171-G177.

30. Mace, OJ, et al. Sweet taste receptors in rat small intestine stimulate glucose absorption through apical GLUT2. *J Physiol* 2007;582(Pt 1):379-392.

31. Pepino, MY and Boume, C. Non-nutritive sweeteners, energy balance, and glucose homeostasis. *Curr Opin Clin Nutr Metab Care* 2011;14:391-395.

32. Depoortere, I. Taste receptors of the gut: emerging roles in health and disease. *Gut* 2014;63:179-190.

Chapter 7: Adverse Effects

1. Chan P, et al. A double-blind placebo-controlled study of the effectiveness and tolerability of oral stevioside in human hypertension. *Br J Clin Pharmacol* 2000;50:215-220.

2. Kimata, H. Anaphylaxis by stevioside in infants with atopic eczema. *Allergy* 2007;62:565-572.

3. Urban, JD, et al. Steviol glycoside safety: are highly purified steviol glycoside sweeteners food allergen? *Food Chem Toxicol* 2015;75:1-8.

Chapter 8: Things You Probably Didn't Know about Stevia

1. Baudier, KM, et al. Erythritol, a non-nutritive sugar alcohol sweetener and the main component of Truvia, is a palatable ingested insecticide. *PLoS One* 2014;9(6):e98949.

2. Hellfritsch, C, et al. Human psychometric and taste receptor responses to steviol glycosides. *J Agric Food Chem* 2012;60:6782-6793.

3. Mari A. Sandell and Paul A.S. Breslin. Variability in a taste-receptor gene determines whether we taste toxins in food. *Current Biology* 2006;16:R792-R794.

4. Lee, RJ, et al. Mouse nasal epithelial innate immune responses to Pseudomonas aeruginosa quorum-sensing molecules require taste signaling components. *Innate Immun* 2014;20:606-617.

Chapter 9: Artificial Sweeteners

1. Long, EL and Haberman, RT. *Review of tumors in rats treated with saccharin and control rats used in study artificial sweeteners 1948-1949.* Institute of Food Technologists: Chicago; 1969.

2. Price, JM, et al. Bladder tumors in rats fed cyclohexylamine or high doses of a mixture of cyclamate and saccharin. *Science* 1970;167(921):1131-2.

3. Sturgeon, SR, et al. Associations between bladder cancer risk factors and tumor stage and grade at diagnosis. *Epidemiology* 1994;5:18-25.

4. *National Toxicology Program (2005).* Toxicity Studies of Acesulfame Potassium (CAS No. 55589-62-3) in FVB/N-TgN(v-Ha-ras)Led (Tg.AC) Hemizygous Mice and Carcinogenicity Studies of Acesulfame Potassium in B6.129-Trp53^{tm1Brd} (N5) Haploinsufficient Mice (Feed Studies). *National Institutes of Health 2005 (NTP GMM-2): 1–113.*

5. Bandyopadhyay, A, et al Genotoxicity testing of low-calorie sweeteners: aspartame, acesulfame-K, and saccharin. *Drug Chem Toxicol* 2008;31:447-457.

6. Zhang, GH, et al. Effects of mother's dietary exposure to acesulfame-K in pregnancy or lactation on the adult offspring's sweet preference. *Chem Senses* 2011;36:763-770.

7. Cong, W, et al. Long-term artificial sweetener acesulfame potassium treatment alters neurometabolic functions in C57BL/6J mice. *PLOS One* 2013;8(8):e70257. doi:10.1371/journal.pone.0070257.

8. Olney, J. Brain damage in infant mice following oral intake of glutamate, aspartate, or cystine. *Nature* 1970;227:609-610.

9. Soffritti, M., et al. First experimental demonstration of the multipotential carcinogenic effects of aspartame administered in the feed to Sprague-Dawley rats. *Environ Health Perspect* 2006;114:379-385.

10. http://dorway.com/aspartame-the-bad-news-repost/peer-reviewed-aspartame-studies/. Accessed 8/4/2016.

11. Abou-Donia, MB, et al. Splenda alters gut microflora and increases intestinal p-glycoprotein and cytochrome p-450 in male rats. *J Toxicol Environ Health Part A* 2008;71:1415-1429.

12. Retting, S., et al. Sucralose causes a concentration dependent metabolic inhibition of the gut flora Bacteroides, B. Fragilis and B. uniformis not observed in the Firmicutes, E. faecalis and C. sordellii (1118.1). *The FASEB Journal* 2014;28:Supplement 1118.1.

13. Qin, X. Sucralose consumption may contribute to inflammatory bowel disease. *World J Gastroenterol* 2012;18:1708-1722.

14. Qin, X. Increased sucralose consumption may explain why Canada has the highest incidence of inflammatory bowel disease in the world. *Can J Gastroenterol* 2011;25:511.

15. Sasaki, YF, et al. The comet assay with 8 mouse organs: results with 39 currently used food additives. *Mutation Research/Genetic Toxicology and Environmental Mutagenesis* 2002;519;103-119.

16. Soffritti, M, et al. Sucralose administered in feed, beginning prenatally through lifespan, induces hematopoietic neoplasias in male Swiss mice. *Int J Occup Environ Health* 2016;22:7-17.

17. Lord, GH and Newberne, PM. Sucralose causes bowel enlargement, kidney mineralization and abnormal pelvic tissue changes in rates. *Food Chem Toxicol* 1990;28:449-455.

18. Goldsmith, LA. Sucralose reduces the weight of spleen and thymus in rats. *Food Chem Toxicol* 2000;38 Supple 2:S53-S69.

19. Patel, RM, et al. Sucralose may be a migraine trigger. *Headache* 2006;46:1303-1304.

20. Otabe. A. et al. Advantame—an overview of the toxicity data. *Food Chem Toxicol* 2011;49:S2-S7.

21. Blaylock, RL. *Excitotoxins: The Taste That Kills*, Health Press: Santa Fe, NM; 1996.

Chapter 10: Sugar Alcohols and Monk Fruit

1. Tamura, M., et al Xylitol affects the intestinal microbiota and metabolism of daidzein in adult male mice. *Int J Mol Sci* 2013;14:23993-234007.

2. Salminen, S., et al. Gut microflora interactions with xylitol in the mouse, rat and man. *Food Chem Toxicol* 1985;23:985-990.

3. Makinen, KK, et al. Thirty-nine-month xylitol chewing-gum programme in initially 8-year-old school children: a feasibility study focusing on mutans streptococci and lactobacilli. *Int Dent J* 2008;58:41-50.

4. Mattila, PT., et al. Effects of a long-term dietary xylitol supplementation on collagen content and fluorescence of the skin in aged rats. *Gerontology* 2005;51:166-169.

5. Mattila, PT., et al. Increased bone volume and bone mineral content in xylitol-fed aged rates. *Gerontology* 2001;47:300-305.

6. Flint, N, et al. Effects of erythritol on endothelial function in patients with type 2 diabetes mellitus: a pilot study. *Acta Diabetol* 2014 Jun;51(3):513-6.

7. No author listed. Xylitol benefits still unproven. *Br Dent J* 2015;218(9):509. doi: 10.1038/sj.bdj.2015.356.

8. Nayak, PA, et al. The effect of xylitol on dental caries and oral flora. *Clin Cosmet Investig Dent* 2014;6:89-94.

9. Milgrom, P., et al. Mutans streptococci dose response to xylitol chewing gum. *J Dent Res* 2006;8:177-181.

10. Roberts, MC., et al. How xylitol-containing products affect cariogenic bacteria. *J Am Denr Assoc* 2002;133(4):435-441.

11. No author listed. Researchers scrutinize xylitol, question its benefits. *Dentistry Today* May 2015, pages 46, 48.

12. Dunayer, EK and Gwaltney-Brant, SM. Acute hepatic failure and coagulopathy associated with xylitol ingestion in eight dogs. *J Am Vet Med Assoc* 2006;229:1113-1117.

13. O'Donnell, S, et al. Non-nutritive polyol sweeteners differ in insecticidal activity when ingested by adult Drosophila melanogaster (Diptera: Drosophilidae). *J Insect Sci* 2016;16(1). pii doi: 10.1093/jisesa/iew031.

14. Matsumoto, K, et al. Minor cucurbitane-glycosides from fruits of Siraitia grosvenori (Cucurbitaceae). *Chem Pharm Bull* 1990;38:2030-2032.

15. Qin, X, et al. Subchronic 90-day oral (Gavage) toxicity study of a Luo Han Guo mogroside extract in dogs. *Food Chem Toxicol* 2006;44:2106–2109.

16. Di, R, et al. Anti-inflammatory activities of mogrosides from Momordica grosvenori in murine macrophages and a murine ear edema model. *J Agr Food Chem* 2011;59:7474–7481.

17. Suzuki, YA., et al. Antidiabetic effect of long-term supplementation with Siraitia grosvenori on the spontaneously diabetic Goto-Kakizaki rat. *Brit J Nutr* 2007;97:770–775.

18. Xiangyang, Q, et al. Effect of a Siraitia grosvenori extract containing mogrosides on the cellular immune system of type 1 diabetes mellitus mice. *Mol Nutr Food Res* 2006;50:732–738.

19. Liu, C, et al. Mogrol represents a novel leukemia therapeutic, via ERK and STAT3 inhibition. *Am J Cancer Res* 2015;5:1308-1318.

Chapter 11: Are Any Sweeteners Safe?

1. http://www.nbs.uk/Livewell/Goodfood/Pages/the-truth-about-artificial-sweeteners.aspx. Accessed 8/1/2016.

2. http://www.gpo.gov/fdsys/pkg/FR-1995-07-20/pdf/95-17505.pdf. Accessed 8/1/2016.

3. https://health.gov/dietaryguidelines/2015/guidelines/. Accessed 9/9/2016.

Index

Dr Fife's Keto Cookery
Nutritious and Delicious Ketogenic Recipes for Healthy Living

A ketogenic diet is one that is very low in carbohydrate and high in fat, with moderate protein. Such a diet shifts the body into an ultra-efficient metabolic state in which fat is utilized as the primary source of fuel in place of glucose (sugar).

This metabolic state, known as nutritional ketosis, has a pronounced therapeutic effect on the body. The diet has proven safe and effective in helping people lose excess weight, improve mental function, balance blood sugar and pressure, improve cholesterol levels, and much more.

Described as the ultimate ketogenic cookbook, Dr. Bruce Fife has compiled into one volume his favorite ketogenic recipes, nearly 450 in all! It includes 70 vegetable recipes, 47 salads and 22 dressings, 60 egg recipes, 50 delicious high-fat sauces for meats and vegetables, as well as a variety of mouthwatering wraps, soups, and casseroles, with a creative array of meat, fish, and poultry dishes. With this resource, you will always have plenty of options to choose from for your daily needs.

No exotic or hard-to-find ingredients here. This is a practical cookbook that can be used every day for life. All of the recipes are simple, with ingredients that are readily available at your local grocery store. None of the recipes include any artificial sweeteners, sugars, flavor enhancers, gluten, grains, or other questionable ingredients. Recipes use only fresh, wholesome, natural foods to guarantee optimal health.

The Coconut Ketogenic Diet
Supercharge Your Metabolism, Revitalize Thyroid Function, and Lose Excess Weight

You can enjoy eating rich, full-fat foods and lose weight without counting calories or suffering from hunger. The secret is a high-fat, ketogenic diet. Our bodies need fat. It's necessary for optimal health. It's also necessary in order to lose weight safely and naturally.

Low-fat diets have been heavily promoted for the past three decades, and as a result, we are fatter now than ever before. Obviously, there is something wrong with the low-fat approach to weight loss. There is a better solution to the obesity epidemic, and that solution is The Coconut Ketogenic Diet. This book exposes many common myths and misconceptions about fats and explains why low-fat diets don't work. It also reveals new, cutting-edge research on the world's only natural low-calorie fat—coconut oil—and how you can use it to boost your energy, stimulate metabolism, improve thyroid function, and lose excess weight.

This revolutionary weight loss program is designed to keep you both slim and healthy using wholesome, natural foods, and the most health-promoting fats. It has proven successful in helping those suffering from obesity, diabetes, heart and circulatory problems, low thyroid function, chronic fatigue, high blood pressure, high cholesterol, and many other conditions.

In this book you will learn:
- Why you need to eat fat to lose fat
- Why you should not eat lean protein without a source of fat
- How to lose weight without feeling hungry or miserable
- How to stop food cravings dead cold
- Which fats promote health and which ones don't (the answers may surprise you)
- How to jumpstart your metabolism
- How to revitalize a sluggish thyroid
- How to use your diet to overcome common health problems
- How to reach your ideal weight and stay there
- Why eating rich, delicious foods can help you lose weight
- Which foods are the real troublemakers and how to avoid them

"This is an excellent book. It is not the usual "diet," where the dieter is limited to certain foods. This weight loss plan actually encourages the consumption of fat on a daily basis...This book is recommended for everyone, with and without a weight problem."
Midwest Book Review

"There is a way of bringing some of your bodily systems into sync with the needs of your body...It is called *The Coconut Ketogenic Diet*, and it will help you in a million little ways, if not a lot of huge ways...to get where you want to be with your body and your weight."
Claudia VanLydegraf, MyShelf.com

Visit Us on the Web

 Piccadilly Books, Ltd.
www.piccadillybooks.com

Stop Alzheimer's Now!
How to Prevent and Reverse Dementia, Parkinson's, ALS, Multiple Sclerosis, and Other Neurodegenerative Disorders

By Bruce Fife, ND
Foreword by Russell L. Blaylock, MD

More than 35 million people have dementia today. Each year 4.6 million new cases occur worldwide—one new case every 7 seconds. Alzheimer's disease is the most common form of dementia. Parkinson's disease, another progressive brain disorder, affects about 4 million people worldwide. Millions more suffer with other neurodegenerative disorders. The number of people affected by these destructive diseases continues to increase every year. Dementia and other forms of neurodegeneration are *not* a part of the normal aging process. The brain is fully capable of functioning normally for a lifetime, regardless of how long a person lives. While aging is a risk factor for neurodegeneration, it is not the cause! Dementia and other neurodegenerative disorders are disease processes that can be prevented and successfully treated.

This book outlines a program using ketone therapy and diet that is backed by decades of medical and clinical research and has proven successful in restoring mental function and improving both brain and overall health. You will learn how to prevent and even reverse symptoms associated with Alzheimer's disease, Parkinson's disease, amyotrophic lateral sclerosis (ALS), multiple sclerosis (MS), Huntington's disease, epilepsy, diabetes, stroke, and various forms of dementia.

The information in this book is useful not only for those who are suffering from neurodegenerative disease but for anyone who wants to be spared from ever encountering these devastating afflictions. These diseases don't just happen overnight. They take years, often decades, to develop. In the case of Alzheimer's disease, approximately 70 percent the brain cells responsible for memory are destroyed *before* symptoms become noticeable.

You *can* stop Alzheimer's and other neurodegenerative diseases before they take over your life. The best time to start is now.

"*Stop Alzheimer's Now!...*will not only be beneficial for Alzheimer's but also for a wide variety of other diseases. I strongly recommend reading this book!"
Sofie Hexeberg, MD, PhD

"A must read for any and all health care professionals, as well as any family members or friends of those stricken by these maladies."
Jeffrey Grill, MD

"A must read for everyone concerned with Alzheimer's disease...the author explains how diet modifications and the addition of coconut oil can drastically change the course of the disease."
Edmond Devroey, MD, The Longevity Institute

"The author's dietary recommendations are a valuable aid to nutritional therapy of chronic neurodegenerative diseases. I recommend this enlightening book to both physicians and those who simply want to better understand how our brain functions."
Igor Bondarenko, MD, PhD

Visit Us on the Web

Piccadilly Books, Ltd.

www.piccadillybooks.com

STOP VISION LOSS NOW!

Prevent and Heal Cataracts, Glaucoma, Macular Degeneration, and Other Common Eye Disorders

Losing your eyesight is a frightening thought. Yet, every five seconds someone in the world goes blind. Most causes of visual impairment are caused by age-related diseases such as cataracts, glaucoma, macular degeneration, and diabetic retinopathy. Modern medicine has no cure for these conditions. Treatment usually involves managing the symptoms and attempting to slow the progression of the disease. In some cases surgery is an option, but there is always the danger of adverse side effects that can damage the eyes even further.

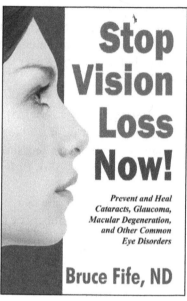

Most chronic progressive eye disorders are considered incurable, hopeless. However, there is a successful treatment. It doesn't involve surgery, drugs, or invasive medical procedures. All that is needed is a proper diet. The key to this diet is coconut, specifically coconut oil and a ketogenic diet. The author used this method to cure is own glaucoma, something standard medical therapy is unable to do.

The coconut based dietary program described in this book has the potential to help prevent and treat many common visual problems including the following:

- Cataracts
- Glaucoma
- Macular degeneration
- Diabetic retinopathy
- Dry eye syndrome
- Sjogren's syndrome
- Optic neuritis
- Irritated eyes
- Conjunctivitis (pink eye)
- Eye disorders related to neurodegenerative disease (Alzheimer's, Parkinson's, stroke, MS)